Guidelines and Standards for
Maternal Serum Screening for Down's Syndrome, Neural Tube Defects and Other Obstetric Problems

Guidelines and Standards for
Maternal Serum Screening for Down's Syndrome, Neural Tube Defects and Other Obstetric Problems

Sharad Gogate
Director
Fetal Medicine Consultancy Services
Surlata Hospital
Mumbai, Maharashtra, India

Forewords
Howard Cuckle
Peter Benn

CBS

CBS Publishers & Distributors Pvt Ltd

New Delhi • Bengaluru • Chennai • Kochi • Mumbai • Pune
Hyderabad • Kolkata • Nagpur • Patna • Vijayawada

Guidelines and Standards for
**Maternal Serum Screening
for Down's Syndrome,
Neural Tube Defects and
Other Obstetric Problems**

ISBN: 978-81-239-2626-1

First Edition: 2015

Published by Satish Kumar Jain and produced by Varun Jain for
CBS Publishers & Distributors Pvt Ltd
4819/XI Prahlad Street, 24 Ansari Road, Daryaganj, New Delhi 110 002, India.
Ph: 23289259, 23266861, 23266867   Website: www.cbspd.com
Fax: 011-23243014   e-mail: delhi@cbspd.com; cbspubs@airtelmail.in.
Corporate Office: 204 FIE, Industrial Area, Patparganj, Delhi 110 092
Ph: 4934 4934   Fax: 4934 4935   e-mail: publishing@cbspd.com; publicity@cbspd.com

Branches

- **Bengaluru:** Seema House 2975, 17th Cross, K.R. Road,
  Banasankari 2nd Stage, Bengaluru 560 070, Karnataka
  Ph: +91-80-26771678/79   Fax: +91-80-26771680   e-mail: bangalore@cbspd.com
- **Chennai:** 7, Subbaraya Street, Shenoy Nagar, Chennai 600 030, Tamil Nadu
  Ph: +91-44-26260666, 26208620   Fax: +91-44-42032115   e-mail: chennai@cbspd.com
- **Kochi:** Ashana House, 39/1904, AM Thomas Road, Valanjambalam,
  Eranakulam 682 018, Kochi, Kerala
  Ph: +91-484-4059061-62-64-65   Fax: +91-484-4059065   e-mail: kochi@cbspd.com
- **Kolkata:** 6/B, Ground Floor, Rameswar Shaw Road, Kolkata-700 014, West Bengal
  Ph: +91-33-22891126, 22891127, 22891128   e-mail: kolkata@cbspd.com
- **Mumbai:** 83-C, Dr E Moses Road, Worli, Mumbai-400018, Maharashtra
  Ph: +91-22-24902340/41   Fax: +91-22-24902342   e-mail: mumbai@cbspd.com
- **Pune:** Bhuruk Prestige, Sr. No. 52/12/2+1+3/2 Narhe, Haveli
  (Near Katraj-Dehu Road Bypass), Pune 411 041, Maharashtra
  Ph: +91-20-64704058/59, 32392277   Fax: +91-20-24300160   e-mail: pune@cbspd.com

Representatives

- **Hyderabad**   0-9885175004
- **Patna**   0-9334159340
- **Nagpur**   0-9021734563
- **Vijayawada**   0-9000660880

Printed at Shree Maitrey Printech Pvt Ltd, Noida

I dedicate this monogram to those brave but unfortunate children with serious genetic defects and birth defects, who brave tremendous odds and still try to live a useful, full life!

Foreword

The possibility of maternal serum screening for birth defects began in the early 1970s with the discovery that second trimester-fetoprotein (AFP) levels were increased, on average, in pregnancies affected by open neural tubes defects (NTDs), principally spina bifida and anencephalus. This marker could be used to identify a small group of women (under 3%) at high risk of these disorders, a group that included four in five open spina bifidas and nine in ten anencephalics. Despite the prospects of reducing the birth prevalence of what was, for some countries, the most common type of abnormality at birth, implementation of screening was slow to be introduced into clinical practice. Concerns were raised about the quality assurance of AFP, an analyte which had at that time not been standardized. Indeed the use of multiples of the median (MoMs) to report results was partly adopted because of vast inter-laboratory differences, in addition to the rapid increase in concentration with gestation. Another concern was that testing might be carried out by commercial laboratories, outside an organized multidisciplinary screening program that provides quality pre-test information and post-test counseling. And many clinicians were concerned that they did not understand the test sufficiently to interpret the results.

Forty years later, this may seem overly cautious given that AFP screening for NTDs is in fact quite simple. But that is a view colored by hindsight after day-to-day experience has familiarized clinicians, and even patients whose friends and possibly mothers have been tested. Indeed there have been many pitfalls on the way. It had to be learnt that the AFP screening test is not diagnostic of itself, but is a way of selecting women for, in the early days, invasive prenatal diagnosis. It needed to be discovered that there are factors associated with false-postive and false-negative results, which can be adjusted for to improve the performance of the test. Situations such as the presence of twins or threatened abortion required special care when interpreting results. And it was necessary to set-up a system of external quality assessment involving pools of serum being divided into aliquots and distributed to laboratories for 'blind' testing. Also software had to be developed to aid epidemiological monitoring of analytic quality allowing screening centers and hospitals to localize and revise important parameters such as the normal median curve.

And that was only the beginning. Raised AFP was shown to also be a marker for other lesions such as open abdominal wall defects. Improvements in ultrasound imaging allowed NTD screening to be carried out by using markers such as 'lemon' and 'banana' signs instead of AFP. Whilst this was simpler it raised a new problem—how to assess the quality of ultrasound—which is more intractable than biochemical marker monitoring. Eventually tertiary ultrasound to visualize the NTD itself was able to replace amniotic fluid biochemistry as the diagnostic test, although small lesions can still be missed. Then came a more profound development, the discovery that AFP was also a marker, albeit weak, of fetal aneuploidy.

The development of Down's syndrome screening using AFP in combination with other maternal serum and ultrasound markers in both the first and second trimesters not only extended the scope of screening but also the complexity. New concepts were required such as risk screening, which makes optimal use of pre-test information such as maternal age and family history, together with the MoM value for each marker. Risk is not only an optimal way of combining all the information it provides a figure that can be used in counseling. Nevertheless, the 'black box' software used to calculate the risk was itself another cause of concern, and a further factor to quality control. Common forms of aneuploidy other than Down's syndrome could also be detected but this required specialist software that was not always available. A late second trimester ultrasound scan to detect structural abnormalities in the fetus has become standard practice in many centers. This too can be used to screen

for aneuploidy using 'soft' markers such as echogenic bowel. However, these results need to be treated cautiously and in light of any screening results earlier in pregnancy; again suitable software is often lacking.

The rate of development in this area does not seem to have abated so that today there is an entirely new way of aneuploidy screening, using maternal plasma cell free DNA. Exactly, as with the early days of AFP screening serious concerns have been expressed regarding commercial testing, quality assessment, counseling, test interpretation and lack of an organized screening framework. Meanwhile, the existing screening tests have been found to have a role in the early prediction of common serious adverse outcomes of pregnancy, such as pre-eclampsia. Here the concerns relate to the next step after screening, namely prevention of the outcome in the group discovered by screening to be at high risk.

These guidelines and standards form an important basis both for existing and emerging screening methods. They provide the educational material needed by clinicians when ordering tests, for managers overseeing programs and for health planners considering implementation of different competing protocols. We can learn from the past 40 years experience to take seriously potential problems, particularly matters of quality assessment and the interpretation of results. Most tests in medicine are carried out in response to symptoms. In contrast, screening tests are pre-emptive being carried out in asymptomatic individuals. Consequently there is a greater ethical imperative to do more good than harm. This publication has an important role to play in this regard too. Dr Gogate is to be congratulated for putting together such a comprehensive document which I recommend to all those with an interest in screening in India.

Howard Cuckle

Columbia University Medical Center
New York, USA

Foreword

The burden of birth defects is considerable from the perspective of the morbidity for affected individuals, the family adjustments required, and the economic consequences for society in general. Some of the most common serious congenital disorders include congenital heart defects, neural tube defects, chromosome abnormalities (most notably Down syndrome) and hemoglobinophathies (including thalassemia and sickle-cell disease). Some birth defects can be entirely prevented through adequate nutrition, diet fortification, and gene carrier identification coupled with preconception genetic counseling. However, many will only be identified during pregnancy or at birth. Consequently, prenatal screening, diagnosis and genetic counseling have become a major component of contemporary obstetric care in most countries.

Laboratory based approaches to prenatal screening for fetal abnormalities began in the early 1970s when it was first proposed that measurement of the concentration of second trimester alpha-fetoprotein in maternal serum could provide a screening test for fetal neural tube defects. In the early 80s it was discovered that alpha-fetoprotein could also be indicative of the presence of fetal chromosome abnormalities. Over the years, additional first and second trimester maternal serum markers have been discovered. Improvements in ultrasound technology have facilitated the recognition of anatomic abnormalities and other 'soft marker' findings associated with aneuploidy. The serum and ultrasound markers are now often combined into sophisticated screening algorithms that facilitate the identification of those pregnancies at highest risk for Down syndrome, additional aneuploidies, other fetal disorders, and pregnancy complications. Maternal co-factors have been identified that affect maternal serum analytes and there has been improvements in the accuracy and performance of many of the laboratory tests. Recently, further improvements in prenatal screening have been made possible through the development of assays that analyze fetal DNA in maternal circulation. In parallel with these advancements in screening, there have been improvements in the technology used in prenatal diagnostics, notably the recent application of chromosome microarrays for the detection of small chromosome copy number variations.

For all of the serum protein, ultrasound markers, and plasma DNA analytes used in screening, the difference between the average affected pregnancy and the average unaffected pregnancy is relatively small. Results are often presented as a patient-specific risk and a small measurement error can often have a surprisingly large effect on the risk estimate. Interpretation of maternal serum markers relies on comparison with referent data sets from unaffected pregnancies. Therefore, the laboratory tests must not drift overtime, not be subject to random measurement errors beyond that present when the referent population data was established, or show errors greater than that seen in the studies establishing the statistical parameters used in the testing algorithm. Ultrasound markers generally require skill in obtaining measurements and also ongoing evaluation of measurements to ensure procedural consistency. Guidelines and standards in screening for fetal disorders and pregnancy complications are therefore extremely important.

In this monograph, Dr Sharad Gogate considers these issues as they relate to screening for fetal Down's syndrome and neural tube defects. The goal is to ensure that obstetricians and laboratory personnel are familiar with requirements necessary for optimal screening performance. Adherence to standards and guidelines should not be viewed as a regulatory burden. Instead, it should be thought of as an inexpensive way to optimize screening. Dr Gogate is to be congratulated for his efforts to promote high standards in prenatal screening and diagnosis.

Peter Benn DSc.
Department of Genetics and Genome Sciences
University of Connecticut Health Center, USA

Preface

The twenty-first century has been dominated by spectacular advances in the field of genetics, rapid advances in imaging moda-lities like Ultrasound, MRI, Bioinformatics. With successful completion of Human Genome project and its fall-outs and now the draft of the Human Metabolome (map of all human body proteins) being prepared, our understanding of human body in health and disease has taken quantum leaps! This has paved the way of genome based preventive medicine with specifically targeted screening, diagnosis and management of various diseases like cancer, metabolic disorders and other ill-understood disorders.

Although the awareness about the importance of genetics in clinical medicine has spread considerably over last few years in medical fraternity, it has not reflected in having a better understanding and confidence about diagnostic tests, screening programs and management of genetic disorders. It was Dr Deepika Deka, Chairperson of the FOGSI, sub-committee of Genetics and Fetal Medicine, who took the initiative to prepare good practice guidelines and handbooks covering the important areas of obstetric ultrasound scans, maternal serum screening for genetic disorders, fetal tissue sampling and laboratory diagnostic testing. Hence, it was a great privilege and pleasure for me when she requested me to prepare a short manual on maternal serum screening for NTD and Down's syndrome for the benefit of practicing obstetricians, laboratory personnel, counselors, imaging experts and the postgraduates.

Though maternal serum screening for Down's syndrome and NTD was initiated in eighties in Europe and USA, in India it was started much later and has been accepted by the general population and clinicians only over last decade or so. It is rather unfortunate that the technical nitty-gritty of this important screening program has not yet been grasped adequately by the practicing obstetricians as well as laboratories providing these screening tests. This results in sub-optimal performance of these tests, ineffective pre- and post-test counseling and lack of standard and proper management of screen positive pregnancies identified by the screening tests.

Over the years the maternal serum screening has acquired many related components like USG screening for aneuploidies, screening for unfavorable outcomes of pregnancies in later months and study of cell-free fetal DNA from maternal blood.

We have tried to cover in this monogram all these aspects about maternal serum screening program in a comprehensive yet practical manner. This should be a very useful reference book for all concerned stakeholders so as to elevate the level of efficiency of this screening program to greater heights and help in down staging of these major public health genetic challenges of aneuploidies, NTD and other unfavorable pregnancy complications.

I sincerely hope this book will find very wide spread readership!

Sharad Gogate

Acknowledgments

I would like to acknowledge all the support and suggestions given as well as the forewords written by Prof Howard Cuckle and Prof Peter Benn.

It is essential for me to thank Dr Hema Purandarey, Dr Pooja Vazirani and Ms Gayatri Jayaraman for sparing valuable time and contribute chapters on their areas of expertise. It has made the monogram more complete and relevant.

I would like to thank Dr Sadanand Talwalkar from Hormone lab, Dr Avinash Phadke and Dr Aparna Jairam from SRL Diagnostics and Dr Phadke Labs for their valuable inputs for the laboratory aspects of maternal serum screening.

It is necessary for me to acknowledge the enthusiastic and timely support from Mr Satish Kumar Jain, Managing Director, and Mr Ramesh Krishnamachari, Regional Manager of CBS Publishers and Distributors.

I would like to thank my wife Prof Alka Gogate for inspiring and supporting me throughout my efforts for preparation of this monogram!

Contents

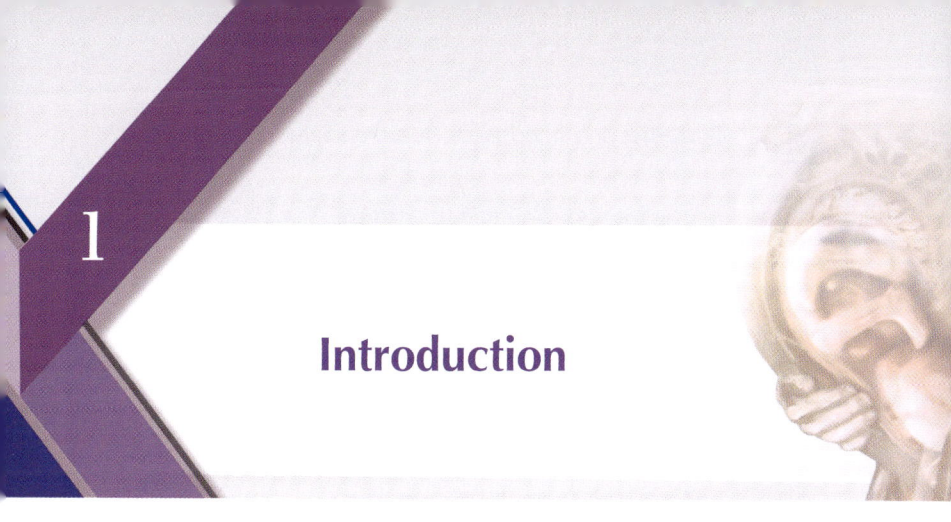

Introduction

All over the world, particularly in developed and developing countries standard of living and health care are improving. Nutritional deficiencies, infections and other preventable causes of morbidity and mortality have been controlled to great extent. Hence non-preventable causes like genetic disorders, congenital anomalies and life style disorders like diabetes, hypertension and coronary artery disease have assumed greater importance as significant cause of morbidity and mortality at all stages of life. With completion of the human genome project our understanding of human genome and its role in human body has undergone sea change. With the recently started project to have a complete draft of all proteins produced in human body, actual expression of genome can be examined in health and in disease.

The Burden of Genetic Disorders and Congenital Malformations

Various measures reflect the population burden of genetic disorders and congenital anomalies. These include the incidence or prevalence of these disorders, associated morbidity and mortality, life expectancy and the social and economic burden on the family and society.

Oocytes and sperms show aneuploidy in 18–19% and 3–5% respectively, as a result 1 in 13 conceptions show chromosomal anomalies. Fifty percent of first trimester abortions are due to chromosomal disorders. Stillbirths and

Table 1.1: The frequency of genetic disorders in 11,69,873 births (source: Baird et al., 1988)

Category	Rate per million live births	% of total births
Autosomal-recessive	1395.4	0.14
Autosomal-dominant	1665.3	0.17
X-linked	532.4	0.05
Chromosomal	1845.4	0.18
Multi-factorial	46582.6	4.64
Genetic unknown	1164.2	0.12
Total	53175.3	5.32
All congenital anomalies	52808.2	5.28

neonatal deaths show chromosomal defects in 5.6–11.5% cases. *Milunsky A., Milunsky J.*[1]

The exact incidence of various categories of genetic disorders is not known, Table 1.1 shows the incidence of these disorders from a very large study by *Baird et al., 1988.*[2]

In a similar study the incidence of genetic disorders in India was found to be 2.3% in a study by *IC Verma*[3].

What is a screening program?

Screening in clinical practice can be defined as use of presumptive methods to identify unrecognized health risk factors or asymptomatic disease in persons/groups that are potentially at elevated risk so as to enable them to benefit from interventions performed before overt symptoms develop. *Braveman PA, Tarimo E.*[4] It is important to emphasize that screening programs should never be considered as a service in itself but should be part of a wider strategy for health promotion, definitive diagnosis and prevention/control of the target disorder in the entire population.

As individual genetic disorders occur at a minute frequency use of diagnostic tests for entire population for such a small risk is not necessary. Genetic screening tests are very useful as we can cover very large population quickly and cost-effectively so as to identify high risk sub-population

Table 1.2: Requirement of a good genetic screening program

- The target disorder should be well-defined, have significant incidence and severity to affect the entire population.
- Screening test should be simple, safe (? non-invasive), affordable, easily available and with acceptable false positivity, specificity and sensitivity.
- Availability of accurate diagnostic test to confirm the diagnosis in screen positive individuals.
- Follow-up management options should be there for the high risk population identified by the screening test.

from the low risk general population. They help reducing need of costly, potentially harmful diagnostic tests; ensure better utilization of costly, labor intensive techniques, save costs while helping the down-staging of the target disease.

At the same time there are certain disadvantages like additional costs and efforts to the patients, problem of false positivity, low sensitivity and specificity, creating anxiety in patients and family, ethical, social issues.

Requirements of a good screening program are shown in Table 1.2.

As of now, none of the screening tests can meet all these criteria!

Problems in Setting up Genetic Screening Program in Developing Countries

Developing, vast countries like India face some unique problems in designing and implementation of genetic screening such as,

- Very large population, lack of awareness, late or no antenatal registration, superstitions/myths.
- Genetic screening not mandatory even in high risk pregnancies, lack of national policy for screening for genetic disorders.
- *Cost factor*: 40% deliveries conducted in public sector, where totally free services are provided. The meagre funds are not adequate to cover even higher priorities like malnutrition, prematurity, infectious diseases.

- 60% deliveries are conducted by private sector, which is self-funded by the patients, as insurance coverage is still very low and most of the insurers do not cover such screening tests.
- Low demand leads to higher cost, which prevents the extensive coverage of population.
- Lack of standard protocols for referral of patients, laboratory assays, interpretation and reporting and lack of proper pre- and post-test counseling.
- Limitations of genetic screening programs: reduced positive predictive value, false positivity, confusing risk assessment, ethical issues.
- Paucity of well-equipped genetic laboratories, particularly in small cities and rural areas, for confirmatory tests, high costs of these tests.

In spite of these problems it is essential for our country to develop national screening programs for genetic disorders with significant incidence, mortality and morbidity. It has been shown that as the economic and health status of a country or major population improves the non-communicable and non-preventable diseases like cancer, metabolic disorders and genetic disorders emerge as important causes of morbidity and mortality.

Some of the main genetic screening programs accepted worlds wide are:

- Down's syndrome
- Neural tube defects
- Hemoglobinopathies like thalassemia, sickle cell disease
- Cystic fibrosis
- Neonatal screening for inborn errors of metabolism.

REFERENCES

1. Milunsky A, Milunsky J. Genetic counseling: preconception, prenatal and perinatal: population study, genetic disorders and the foetus. Editor: Milunsky A; John Hopkins. Fifth edition page 1, 2004.

2. Baird PA, Anderson TW, Newcomb HB, et al. Genetic disorders in children and young adults: Am J Hum Genet 1988; 42:677.

3. Verma IC. Burden of genetic disorders in India. Indian J Paediatr 2001; 67:893.

4. Braveman PA, Tarimo E. Screening in primary healthcare; WHO publication; page 6, 1994.

2

Epidemiology of NTD and Down's Syndrome

Neural Tube Defects (NTD) are one of the most common birth defects affecting the human fetus with significant morbidity and mortality. An important distinction from the point of view of prenatal diagnosis is that between 'open' and 'closed' spina bifida. Open in this context means that there is some exposure of neural tissue or the lesion is completely covered by a thin transparent membrane, and closed means covered by skin or a thick opaque membrane. Open spina bifida can be more readily diagnosed by biochemical tests and has worse prognosis. About one-third of infants with open lesions survive to 5 years and most survivors have severe handicap due to hydrocephalus, urinary and anal incontinence and paralysis of the lower limbs. For closed lesions, about two-thirds survive to 5 years and one-third of survivors have severe handicap. About one in six spina bifida lesions are open. The incidence of NTD in India varies from 1.7 to 12.2/1000 live births. *BDR News.*[1] These include anencephaly, open and closed neural tube defects.

Down's syndrome (trisomy 21): Trisomy 21 is the most common autosomal aneuploidy with significant impact on the fetal and neonatal well-being, burden on the families and society in form of moderate to severe mental retardation and variable expression of several malformations such as congenital heart disease, hypothyroidism as well as predisposition to certain cancers like leukemia. Morbidity

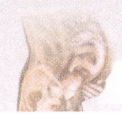
has been improved in recent decades through more effective treatment of associated cardiac, digestive and respiratory symptoms. However, in addition to the mental handicap many affected adults may experience cognitive deficits due to pathological changes in the brain normally associated with Alzheimer's disease. In two large population studies involving over 100000 live births the incidence of trisomy 21 was 1.2–1.7 per 1000 live births. *Lubs HA, Ruddle FH*[2] *and Nielsen J, Wohlert M.*[3] There is no national data on incidence of Down's syndrome from India, but in a large multi-centric study by *Verma IC*[4] it was 1:1139 live births.

Hence, these were the two disorders targeted for developing screening and diagnostic programs in pregnant women. *Over the last 5–6 years the maternal serum screening test has evolved with more serum, USG markers and is useful for screening for many more disorders, late onset complications like PIH, IUGR, preterm labor, etc.*

Basic Concept

Basic concept of Down's and NTD screening is that large number of molecules are produced by fetus or placenta which cross the feto-placental barrier into the maternal circulations. These are potential markers for screening from maternal blood.

Each such marker is evaluated for its distribution on a Gaussian curve in affected and normal pregnancies. By using detailed statistical analysis, such as regression analysis, Mahalanobis distance etc. and evaluation of individual large studies and meta-analysis of many such studies, the suitability of such markers is decided. Parameters for evaluation include detection rate (DR), false positivity (FP) and negativity (FN) and odds of being affected with a positive result (OAPR).

As all the markers show variation as pregnancy advances, gestational age specific median values have to be generated for each week/day of pregnancy. Absolute values are not useful, multiples of median values (MoM) have to be calculated and converted to log 10 values. For open neural tube defect screening, alpha fetoprotein is used. For Down's syndrome screening, no single marker has the desired sensitivity and specificity hence more than one marker, along with maternal age, are considered in a multi-variate analysis. By using special statistical software. The risk of recurrence is calculated by modifying the priori risk based on maternal age, family history of aneuploidy by a likelihood ratio derived from individual marker profile. This risk is compared to the fixed cut-off risk to give final screening status. *Wald N J et al.*[5]

By evaluation of a large number of such molecules certain markers have been identified, the most promising and widely used ones are:

- Alpha fetoprotein (AFP)
- hCG (total hCG)
- free beta-hCG
- uE3
- PAPP-A
- Inhibin-A

Some more markers have been tested but not yet found significant in prospective clinical trials such as:

- ADAM-12
- Invasive trophoblast antigen (ITA)

REFERENCES

1. Proceedings of the BDR meeting. BDR News. Ed. Foetal Care Research Foundation; Chennai, India. 2005; 5:1.
2. Lubs HA, Ruddle FH. Chromosomal abnormalities in human population: estimation of rates based on New Haven Newborn Study. Science 1970; 169:495

3. Nielsen J, Wohlert M. Chromosomal abnormalities found among 34910 newborn children: results from Denmark Study. Hum Genet 1991;87:81.

4. Verma IC. Burden of genetic disorders in India. Indian J Paediatr 2001; 67:893

5. Wald N J, Kennard A, Hackshaw A, McGuire A. Antenatal screening for Down's syndrome. J of Medical Screening; 1997; 4:181–246.

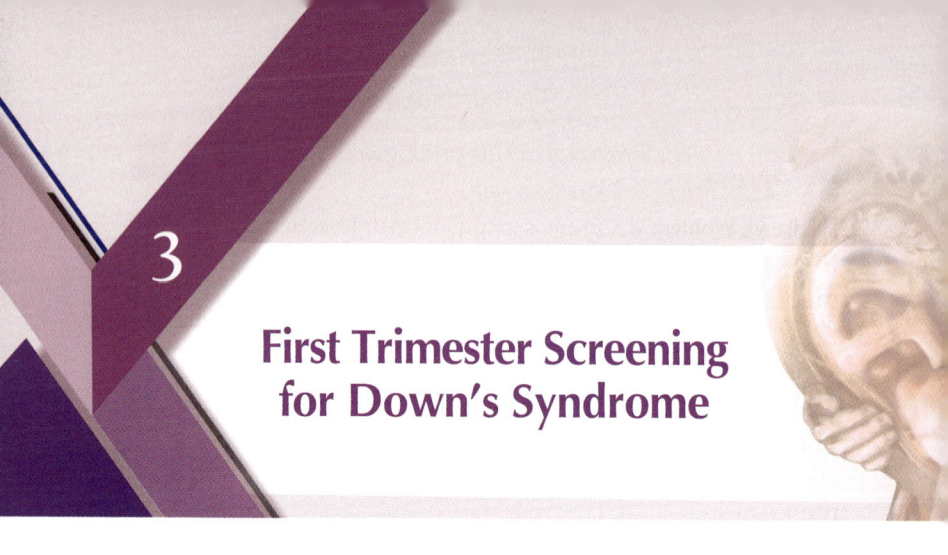

3

First Trimester Screening for Down's Syndrome

First trimester screening earlier in pregnancy offers substantial benefits. Firstly, there is early detection and option of first termination of pregnancy (if so indicated and opted by patient). It can be done with less psychological trauma and less risk to the mother's health. Secondly, there is early assurance following a negative screening result. Moreover, screening in the first trimester is more efficacious than in the second.

Initially, first trimester policy was based on ultrasound NT alone, at 11–14 weeks (CRL 45–84 mm), but now it is combined with maternal serum PAPP-A and free hCG at 11–14 weeks or hCG at 12–13 weeks. Meta-analysis shows that, NT alone at 11 weeks has a detection rate 7–14% greater than the best second trimester serum combination. Adding serum PAPP-A and free hCG at 10 weeks to NT increased the detection advantage over the best second trimester serum combination to 16–25%. The advantage of NT alone or in combination was less at later gestations and by 12 weeks substituting hCG for free β-hCG made little difference. *Nicolaides KH, SabireNJ, Snijder RJM.*[1]

To set-up such a Program it is Essential to go in a Systematic Manner

1. The entire team of local clinicians, paramedical workers and the laboratory personnel should understand the basic concept of first trimester screening.

2. Evolve pre- and post-screening counseling, prepare patient information brochures and various proforma like informed consent, test report and interpretation of the test.
3. It is also vital to collect adequate baseline data for the screening parameters like the Nuchal translucency, PAPP-A and free β-hCG in ongoing pregnancies.
4. The statistical analysis of these data should be evaluated and compared with other previous studies to ensure that the local median values of all the test markers are robust and truly representative of the local population. Only after these have been established then one can start screening the population. *Gogate SG.*[2]

It is essential to have strict quality control at all stages of the program and regular internal as well as external audit should be done to monitor the program.

NUCHAL TRANSLUCENCY (NT) MEASUREMENT

Background

There is an agreed policy defining screening for Down's syndrome in England, which describes the possible use of NT as part of a screening method. The following standards for the measurement of NT have been produced to support the agreed policy. They should be periodically reviewed and updated where necessary.

Standards

1. There must be a local policy and protocol for all departments using NT screening. Each center must take part in the audit and monitoring program.
2. All ultrasound practitioners performing NT measurements must be appropriately trained and accredited and their results subjected to rigorous valid audit and performance management as approved by the regional advisory body. To assure continuing satisfactory performance each ultrasound practitioner must perform a minimum of 50 nuchal translucency measurements per year.

3. The ultrasound equipment must have a cine-loop function and the calipers must have a precision to one decimal point, i.e. 0.1 mm. Harmonic imaging could be helpful to optimise the image and procure the right section but should be turned off when making a measurement. All equipment should display TI (Thermal Index) and MI (Mechanical Index) in accordance with the output display standard.

4. Ultrasound scanning equipment must meet the European Council directive, enforced by the medicines and health care regulatory agency, to ensure that it is safe and effective to use. Equipment should be repaired and maintained in accordance with manufacturer specifications with particular reference to caliper accuracy.

5. The crown rump length (CRL) must be measured according to the approved national guidelines. The CRL should be in the range 45–84 mm. The measured image of the CRL must be retained in the patient record and communicated to the laboratory along with sample.

6. NT can be measured successfully by transabdominal ultrasound examination in approximately 95% of cases; in the others, it may be helpful to perform transvaginal sonography. It may be appropriate to recall the woman for a second scan if the NT cannot be measured due to inappropriate position of the fetus. However, if there is failure to obtain an NT measurement then the woman must be offered a biochemical screen without NT.

7. A midline sagittal section of the fetus should be obtained with the fetus horizontal on the screen, supine (Fig. 3.1). The fetus should be in the neutral position with the head in line with the spine, neither hyper extended nor flexed. The NT should be measured once the magnification of the fetus is made as large as possible before the image is selected for the measurement of NT. To improve the accuracy of the NT measurements only the fetal head and shoulders should be visible on screen (Fig. 3.2). Small

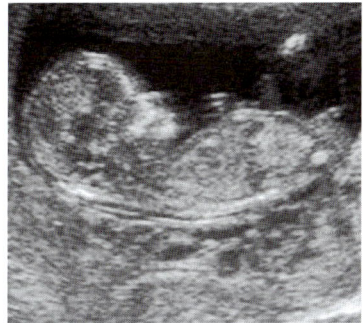

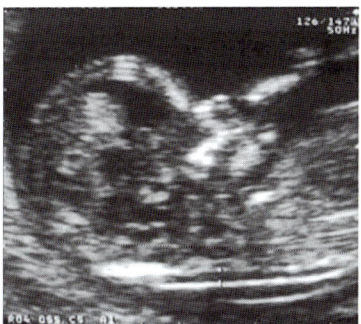

Fig. 3.1: USG image (incorrect)   **Fig. 3.2:** USG image (correct)

movement of the calipers on the magnified image must produce only a 0.1 mm change in the measurement. It may be helpful to reduce the gain to improve the image quality.

8. Care must be taken to distinguish between fetal skin and amnion because, at this gestation, both structures appear as thin membranes. This is achieved by waiting for spontaneous fetal movement away from the amniotic membrane; alternatively, the fetus is moved off the amnion by asking the mother to cough and/or by tapping the maternal abdomen.

9. The maximum thickness of the subcutaneous translucency between the skin and the soft tissue overlying the cervical spine should be measured (Fig. 3.2).

10. The caliper selected should be a vertical cross. Measurements should be taken with the horizontal lines of the calipers placed 'on' the lines that define the nuchal translucency thickness, not in the line and not in the translucency (Fig. 3.3).

11. The image should be unfrozen and the measurement repeated until the optimum image, meeting the above criteria has been obtained. This image must be retained in the patient record and communicated to the laboratory along with sample.

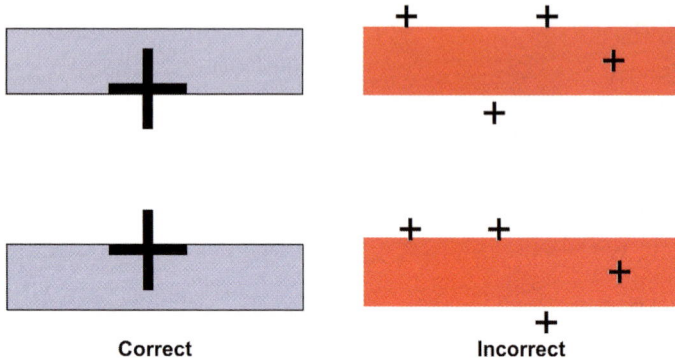

Correct   **Incorrect**

Fig. 3.3: Placement of calibers for NT measurement

12. A risk calculation software package which complies with the European Union Directive of 2003 on CE marking must be used. The calculation package should be comply with the recommendations of the national screening committee's specification document. *Nicolaides KH et al.*[3]

13. Median NT graph (Fig. 3.4).

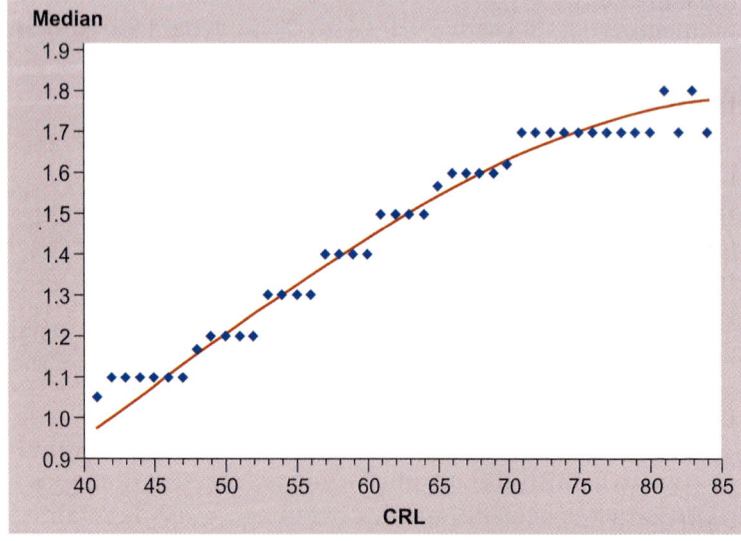

Fig. 3.4: Median NT (mm) according to CRL (mm): observed and regressed values

MATERNAL SERUM SCREENING

Trisomy 21 risk evaluation in 1st trimester pregnancies using maternal serum markers:

In 1st trimester combined screening, we evaluate the risk between 10 weeks 6 days and 13 weeks 6 days gestation. Parameters for 1st trimester test are PAPP-A and free β-hCG along with NT.

Brief Details on Individual Parameters

PAPP-A, i.e. pregnancy associated placental protein A is a glycoprotein which is produced in high concentration by the trophoblasts during pregnancy and released into maternal circulation.

The levels of this protein rise steadily with gestational age, most noticeably during the last part of pregnancy. Reduced levels of PAPP-A are associated with Down's syndrome. In a study of 210 trisomy 21 pregnancies the median PAPP-A at 11–13th week was 0.51 MoM (95% CI was 0.44–0.56). It's measurement during 10–13th week gestation is reported to have significant utility in screening for Down's syndrome and other chromosomal anomalies. Detection rate (DR) of this molecule alone is only 40%.

Free β-hCG again a glycoprotein hormone is present in blood circulation during pregnancy in very low concentration, about 1% of total hCG. However, it is unstable at room temperature. It is secreted by the placental tissue almost from the time of implantation and serves to support the corpus luteum in early pregnancy. Its assessment is reported to have significant utility in 1st trimester screening for Down's syndrome in 11–13th week of pregnancy. The levels rise in the first trimester and then subsequently decrease as the gestation advances. The levels of this molecule are increased in trisomy 21 pregnancies. In a study of trisomy 21 pregnancies median free-beta-hCG was 2–15 MoM (95% CI 1.94–2.33). DR with this marker alone is only 35%. *Spencer K et al.*[4]

Power of first trimester screening: When tested together these two have DR of 60% at 5% FPR, when combined with NT study the DR goes up to 85–87%. Additional inclusion of nasal bone, tricuspid regurge and ductus venous Doppler flow increases the DR to over 92–94%.

Sample Processing

The blood sample collected, at the appropriate gestational age as determined from ultrasound, is expected to be stored in refrigerator till dispatch to the laboratory on dry ice. If a facility for serum separation is available at the collection unit, the sample can be processed within 30 min to separate the serum which in turn can then be stored frozen till delivery to the lab. Complete details of the patient as outlined below, must be supplied to the lab for an accurate evaluation.

Assay Methods

PAPP-A and Fr β-hCG, most of the labs all over are currently using chemiluminescence methods, either by MAIA (magnetic particle separation method) (e.g. delphia) or by alkaline phosphatase separation,which are available on fully automated analysers. Radioimmunoassays (RIA) based methods were used in past but now not in use.

Laboratory Standards

All laboratories need to include control materials which reflect clinically relevant concentrations in each assay run. Control materials need to have defined clinical ranges for acceptance of the run. Control data should be carefully evaluated to look for assay drift or other evidence of change in test performance. Changes in kit lots should be evaluated by comparing the old lot and new lot, prior to placing the new lot in service. MoMs for each analyte should be reviewed at least monthly and have values within the range 0.95–1.05. Regression curves for the median concentrations against gestational age should also be regularly reviewed to ensure conversion of concentrations into MoM values are accurate.

Pre- and Post-test Counseling

This is one of the most important parts of the aneuploidy screening programs. (Although, it is also the most neglected in present scenario!) Aneuploidy risk assessment is a component of a broad set of prenatal clinical services that should be offered from 10 to 14 weeks gestational age whenever possible, and not a stand alone activity. Services can include genetic counseling, screening for pregnancy complications and other fetal conditions, diagnostic testing (chromosome analysis, microarray analysis, fetal cell-free DNA from maternal blood and other genetic testing), midwifery and obstetrical interventions in high risk patients identified. For women who only come into care after the first trimester, risk assessment testing should be made available as soon as possible.

Ideally this counseling should be provided by genetic counsellors or by clinicians or fetal medicine consultants, unfortunately in actual practice this is not possible as there is non-availability of genetic counsellors, obstetricians are too busy and are not fully oriented to offer effective counseling. There is need to train the hospital paramedical personnel like nurses in basics of this counseling. This lack of effective counseling creates lot of confusion, anxiety as well as higher chances unnecessary interventional procedures.

Pre-test Counseling

Prior to undergoing prenatal screening, pregnant woman and her spouse/family should be given information on the principles of screening process, it is advantages and disadvantages, details of the target clinical conditions to be screened and it's impact on the affected individual as well as the family. The patient should be explained the process of the screening test, its cost, sensitivity, specificity also needs to be discussed. The likely outcomes of the screening test and the implications should be explained clearly. The pregnant woman and her family should be provided with an opportunity to discuss this with a health professional, as well as family members before making a personal decision to accept or decline screening.

Post-test Counseling

Once the screening test reports are ready, it is necessary to explain and interpret the test results to the patients and family. The sensitivity, specificity, etc. of the test should be re-iterated. The difference between screening test and diagnostic test should be stressed again. In case of clear cut screen negative report, patient should be reassured. While in borderline and screen positive results, need for invasive diagnostic tests like chorionic villous sampling, amnio-centesis should be stressed. Detailed counseling for this should be provided by the fetal medicine consultant per-forming these tests. For those not ready to accept the invasive testing noninvasive cell-free fetal DNA test (NIFT) or follow-up screening like second trimester quadruple test, anomaly detection USG scan should be offered and discussed. The counseling should be non-directional, unbiased and sensitive to ethical and religious views of the family and patient.

Clinical requirements are, accurate NT measurement, CRL measurement, GA on ultrasound date (*it is necessary to send USG images of CRL and NT measurements*), patient's date of birth, the current body weight, pregnancy status- single or twin pregnancy and any confounding factors like history of chromosomal anomalies, bleeding during pregnancy, diabetes and whether the conception is by ART like IVF/ ICSI. If donor eggs are used, need donor woman's age, relevant history. In case of surrogacy, genetic mother's age, relevant information should be given. It is necessary to have USG images of NT and CRL with the blood sample. Ideally the USG scan and blood collection should be done on same day, definitely not more than 2–3 days apart.

All of these details must be supplied to the laboratory for accuracy of risk evaluation. The data obtained from the parameters as above is entered in the suitable software along with the patient details. The software calculates the statistical value; medians and MoM values based on the entire details and evaluate the risk for Down's syndrome. As there is a significant variation in the median values of the test markers in different ethnic and geographic groups it is essential to have baseline data from the local community to give proper risk assessment.

First Trimester Screening Report and Interpretation

The test report should be comprehensive, easy to understand, there should be an explanatory note giving clarification of the test result, accuracy and limitations of the screening test, what are the post-test options and about the expected follow-up. This will help the referring physician to offer post-test counseling and further management.

Screen positive (risk ratio 1:100 or higher): Diagnostic testing should be recommended in all screen positive pregnancies. Fetal karyotyping by chorionic biopsy is the logical choice, but the patient should have freedom to choose the fetal tissue sampling procedure. There is a lot of bias in clinicians about the safety, accuracy of CVS vis-à-vis amniocentesis. Although it has been proved abundantly that CVS, if properly performed by trained personnel, as per standard protocol, is as safe and reliable as amniocentesis. As for those patients refusing fetal tissue sampling, anomaly scan at 18–20 weeks, which includes detailed review of fetal, placental anatomy for structural defects, ultrasound markers for aneuploidies and second trimester screen test should be offered to. It is also essential to counsel the patient of the option of NIFTY (Advanced non-invasive fetal cell free-DNA testing for aneuploidy.

Screen negative (risk ratio <1:1000): Post-test counseling should be offered along with recommendations for anomaly scan along with fetal 2-D echo at 20 weeks of pregnancy.

Borderline reports (risk ratio >1:100 to <1:1000): Counseling, second trimester screening or diagnostic test in view of positive history or USG scan findings (refer to report and explanatory note proforma in the annexure)

Follow-up

All women with large NT (even if they are screen-negative) should receive second trimester ultrasound to rule out a cardiac abnormality or other fetal defect. All women with continuing pregnancies should receive either the AFP test or second trimester ultrasound to rule out a NTD. Very low levels of PAPP-A (<0.4 MoM) should be monitored by Doppler studies of uterine arteries at 20 wks to screen for PIH, IUGR, etc. in third trimester.

It may be beneficial to start prophylactic low dose aspirin therapy in patients with high uterine artery resistance detected at 11–14 weeks, very low MoM PAPP-A.

It is essential to instruct the patient to give follow-up information about the course of pregnancy, final outcome and post-delivery corroboration of the test result.

REFERENCES

1. Nicolaides KH, SebireNj, Snijders JM. Nuchal translucency and maternal serum biochemistry. The 11–14 weeks scan. Parthenon Publishing. 1999; pp. 38–41.

2. Gogate SG. Preventive Genetics-Holistic healthcare; Preventive Genetics, Gogate SG, (Editor). Jaypee Brothers, New Delhi, 2006; pp. 3–24.

3. Nicolaides HN, Sebire J, Snijders JM, (Editor). The 11–14 week scan. The diagnosis of fetal abnormalities. 1999. pp. 1467–2162.

4. Spencer K, Souter V, Tul N, Snijder R, Nicolaides KH. A screening program for Trisomy 21 at 10–14 weeks using foetal nuchal translucency, maternal serum free β-hCG and PAPP-A. Ultrasound Obstet Gynecol 1999; 13: 231–7.

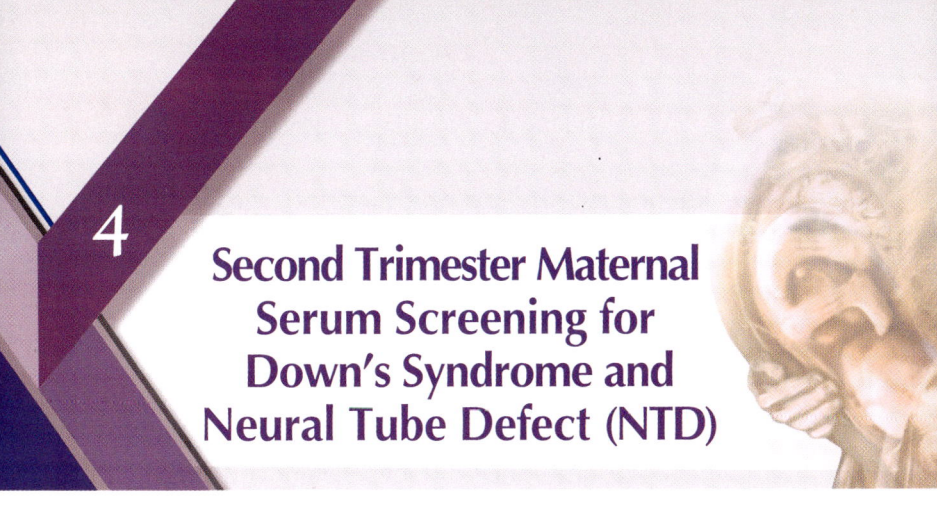

4

Second Trimester Maternal Serum Screening for Down's Syndrome and Neural Tube Defect (NTD)

DOWN'S SYNDROME SCREENING

Neural tube defects are among the most common birth defects affecting the human fetus with significant morbidity and mortality, also trisomy 21 is the most common aneuploidy with significant impact on the fetal and neonatal wellbeing in form of moderate to severe mental retardation and several genetic deformities. Hence these are the two disorders targeted for developing screening and diagnostic programs in pregnant women. The three marker screening along with maternal age was initiated in 1988.

The emphasis in antenatal screening has been to develop earlier, more reliable, cost effective and safer screening tests so as to cover maximum low risk population to identify the high risk sub-group. As a result diagnostic testing like fetal tissue sampling and confirmatory tests (which are potentially harmful, labor intensive and expensive) can be offered to a smaller subpopulation with maximum detection of affected fetuses.

The markers used commonly in second trimester screening are:

- Alpha fetoprotein (AFP)
- Total human chorionic gonadotropin (total hCG)
- Unconjugated estriol (uE3)
- Inhibin-A

First three markers constitute the "triple marker test", addition of inhibin-A makes it "quadruple test".

AFP or alpha fetoprotein is produced in large amounts by the fetus and only in trace amount by the mother. The fetal concentration of this glycoprotein hormone rapidly decreases towards birth. The amniotic fluid also shows presence of AFP which is about 100 fold lower than in fetal serum but approx 1000 times higher than the maternal serum concentration. The levels in maternal serum steadily rise till about 22–24 wks gestation and then remain at plateau till full term. The amniotic fluid levels on the other hand fall with advancing gestation, from 12 weeks till 24 weeks. Elevated maternal serum AFP levels are often seen when multiple pregnancies or a NTD is present whereas the levels remain low when fetal Down's syndrome is present. AFP is an important and useful marker for screening during the second trimester and was the first to be used for screening. AFP is most informative ideally between 16 and 18 weeks. For Down's syndrome screening, AFP along with maternal age has a DR of 36% at 5% FPR. *Milunsky A, Canick JA.*[1]

Unconjugated estriol (uE3) is synthesized in placenta from fetal precursors. It is highly specific for pregnancy and reflects a better assay of feto-placental production of estriol than total estriol assay. The levels increase throughout gestation from 10 weeks till full term. Persistently low or rapidly falling estriol levels suggest fetal distress. Its serial determination is useful in monitoring complicated pregnancies such as those with diabetes, IUGR, hypertension etc. For Down's syndrome screening uE3 along with maternal age has DR of 41% at 5% FPR. *Canick JA et al.*[2]

Total or intact hCG as a marker in 2nd trimester risk evaluation has been well accepted. Human chorionic gonadotropins are glycoproteins with epitopes, hCG has two non-identical sub-units alpha and beta. The levels rapidly rise during 1st trimester, steadily decline during the 2nd trimester, and then remain lower during the 3rd trimester. Both total hCG and free beta-hCG levels are increased in maternal serum in Down's syndrome conceptions. For Down syndrome screening, total hCG has DR of 49%. *Bogart et al.*[3]

Inhibin-A is a protein dimmer with alpha and beta sub-units. Only inhibin-A is seen in pregnant blood samples. Its levels are increased in Down's pregnancies but not before 13 weeks of gestation. For Down's syndrome screening, inhibin-A has DR of 44%. *Van Lith et al et al.*[4]

Clinical requirements: Date of birth, LMP, BPD measurement, GA on ultrasound date, the current body weight and pregnancy status — single or twin pregnancy.

The accuracy of evaluation from the software is enhanced if most of these details are supplied to the laboratory. The data obtained from the biochemical marker tests is entered along with the patient details. The software calculates the statistical value medians and (MoM) based on the entire details fed and evaluates the risk for Down's/neural tube defects.

Sample Processing

The blood sample collected at the appropriate GA as determined from USG, is expected to be stored in refrigerator till dispatch to the laboratory. If a facility for serum separation is available at the collection unit, the sample can be processed within 30 min to separate the serum which in turn can then be stored frozen till delivery to the lab. As freezing affects uE3 assays, it is better not to freeze the serum. Complete details of the patient as outlined above in the beginning must be supplied to the lab for an accurate risk assessment.

Efficacy of Second Trimester Serum Screening

Detection rate of triple markers along with maternal age is 60% at FPR of 5%, addition of inhibin-A it increases to 65–67%. Results can be combined second trimester ultrasound markers to increase the detection rate and/or lower the false-positive rate.

The screening performance of the commonly used markers for Down's syndrome is shown in Table 4.1.

Table 4.1: Second trimester: Down's syndrome detection rate for a given false-positive rate. *Cuckle H*[5]

Marker combination	False-positive rate		
	1%	*3%*	*5%*
AFP and free β-hCG	38%	53%	61%
AFP, free β-hCG and uE₃	42%	58%	65%
AFP, free β-hCG, uE₃ and inhibin	50%	64%	71%
AFP and hCG	34%	48%	56%
AFP, hCG and uE₃	39%	53%	60%
AFP, hCG, uE₃ and inhibin	47%	60%	67%
Plus ultrasound NF			
AFP and free β-hCG	60%	72%	77%
AFP, free β-hCG and uE₃	63%	74%	79%
AFP, free β-hCG, uE₃ and inhibin	68%	78%	83%
AFP and hCG	57%	68%	74%
AFP, hCG and uE₃	60%	71%	76%
AFP, hCG, uE₃ and inhibin	65%	75%	80%

In order to increase the power of the screening program various confounding factors which have direct or indirect bearing on the screening performance have to be identified and evaluated. Such factors include,

- Maternal weight
- Multiple pregnancies
- Previous affected pregnancies
- Use of ART techniques
- Smoking
- Ethnicity
- Maternal disorders like insulin dependent diabetes, chronic renal failure.

It has been observed that frequently after a properly done first trimester combined test (NT, serum markers PAPP-A and free beta-hCG), clinicians often advice the patients to undergo second trimester triple/quadruple tests. This is illogical as the DR of second trimester tests is much less than

first trimester combined test, this also gives a second risk factor, which is confusing. Most importantly the advantage of early screening, diagnostic testing by CVS is lost and patient gets pushed into second trimester diagnosis by amniocentesis.

There are many software available for the risk assessment, we have been using alpha plus software, procured from UK and is developed by Prof NJ Wald. We have provided our own data from over 200 samples tested by us and their statistical data has been incorporated in our alpha software to minimize all variations arising out of ethenic and genetic patterns. Thus the false positivity/negativity rate is maintained to 3% or less. Other labs have used different software platforms for computation of risk assessment, which may or may not have these modifications.

MATERNAL SERUM SCREENING FOR NTD

Elevated levels of maternal serum AFP in NTD, e.g. anencephaly was first noted in 1971 by Hino et al. this led to MSAFP screening of pregnant women in USA and England in 1973.

Population screening for NTD using maternal serum AFP and ultrasonography scans coupled with peri-conceptional folic acid fortification/supplementation is one of major success stories in preventive genetics. As a result of this program incidence of anencephaly and open spina bifida declined in England and Wales by 95% (from 1960s 4 per thousand to 0.2 per thousand births in 1990s). *Morris JK, Wald NJ.*[6]

As maternal serum levels rise by 15% per week in the second trimester it is necessary to have gestational age-specific median values. The current practice is to have day specific median values to calculate the MoM values. Optimum dating of the pregnancy uses a first trimester dating scan. *Knight GJ.*[7]

The optimum window for efficient MSAFP screening is between 16–18th week, not before 15th week and not later than 20th week. Most laboratories use >2.0 MoM or >2.5 MoM as upper limit cut-off for normal range. Assays of MSAFP are done either by radioimmunoassay or chemiluminescence or immunofluorescence methods. It is essential for each laboratory to have it's own normal range and have continuous updating of these values with monitoring through quality control.

Confounding factors in NTD screening:
- Gestational age
- Maternal weight
- Maternal ethnicity
- Maternal insulin dependant diabetes
- Multiple pregnancies
- Ante-partum bleeding/interventional procedures like CVS, amniocentesis

An elevated MSAFP should be rechecked and reconfirmed; this should be followed by a level two anomaly scan to look for NTD as well as other structural anomalies. With advances in high resolution ultrasound scans, over 95% NTDs can be detected with <1% false positivity. If elevated amniotic fluid AFP levels are found, acetylcholinesterase (AChE) testing is done by polyacrylamide gel electrophoresis (PAGE) on amniotic fluid. This test can rule out non-neurological lesions. Chromosomal studies of the amniotic fluid cells can look for chromosomal defects associated with NTD.

Amniotic fluid levels of AFP decrease by 15% per week between 15–22nd weeks hence it is essential to have week specific values expressed as MoM or standard deviations for evaluation. Fetal/maternal blood contamination can give elevated MSAFP levels.

Other causes of elevated MSAFP are:
- Fetal demise
- Low birth weight

- Ventral wall defects (gastroshisis and omphalocele)
- Congenital nephrosis
- Mild benign obstructive uropathy
- Hereditary elevated MSAFP

Role of Second-trimester Ultrasound in Screening for Down's Syndrome

In recent years, the focus of antenatal screening for Down syndrome has shifted from the second to the first trimester of pregnancy in most developed countries. Where does that leave second-trimester ultrasound markers of Down syndrome? These have four potential roles: in women with a high risk following first-trimester screening; in those with a low risk following such screening; in twin, and in all pregnancies in which first trimester screening was not performed. It is not uncommon for a woman who has had a positive combined first trimester test result to delay a decision over invasive prenatal diagnosis until ultrasound evidence for/or against Down syndrome can be found at 14–24 weeks. Under such situations second trimester USG markers may be used to modify the risk shown by serum screening tests a simplistic interpretation, in which the presence of one or more markers is taken to be sufficient to tip the balance in favor of invasive testing and the absence of any markers is sufficient to contraindicate testing, is no longer acceptable. Instead, information from the scan should be used to revise the risk from the combined test: when a marker is present the risk will be increased and when it is absent the risk will be reduced, but the magnitude of these changes differs from marker to marker. (*H. Cuckle, R. Maymon*).

In a recent meta-analytical study *Agathokleous et al.* have done meta-analysis of second-trimester markers for trisomy 21 by pooling data from over 48 studies of ultrasound markers for trisomy 21 (Table 4.2). The most widely examined markers in these studies are lateral cerebral ventriculomegaly (10 mm or more), absent or hypoplastic nasal bone, increased nuchal fold thickness (6 mm or more), intra-cardiac hyperechogenic focus, aberrant right subclavian artery (ARSA), hyperechogenic bowel (echogenicity equal to that of bone), mild hydronephrosis (4 mm or more) and shortening of the femur or humerus based on a cut-off of the respective bone length as a function of gestational age or biparietal diameter.

Table 4.2: Pooled estimates of detection rate (DR), false positive rate (FPR) (both at 95%) and positive and negative likelihood ratios (LR+, LR–) and of sonographic markers for trisomy 21 and estimated likelihood ratio (LR) of individual isolated markers

Marker	DR(%)	FPR(%)	LR+	LR–	Isolated marker LR
Increased NF	26.0	1.0	23.30	0.80	3.79
Hypoplastic NB	59.8	2.8	23.27	0.46	6.58
Ventriculomegaly	7.5	0.2	27.52	0.94	3.81
ARSA	30.7	1.5	21.48	0.71	3.94
Echogenic cardiac focus	24.4	3.9	5.83	0.8	0.95
Echogenic bowel	16.7	1.1	11.44	0.9	1.65
Short femur/humerus	27.7/30.3	6.9/4.6	3.7/4.8	0.8/0.74	0.6/0.7

In this meta-analysis the combined negative LR of all markers, including short femur but not short humerus, was 0.13, implying that if a systematic ultrasound examination is carried out and all markers are excluded there is a 7.7-fold reduction in risk. The clinical implications of these findings are that firstly, if a systematic second-trimester ultrasound examination demonstrates the absence of all major defects and markers there is a 7.7-fold reduction in risk for trisomy 21; secondly, the detection of any one of the markers during the scan should stimulate the sonographer to look for all other markers or defects; thirdly, the post-test odds for trisomy 21 is derived by multiplying the pre-test odds by the positive LR for each detected marker and the negative LR for each marker demonstrated to be absent; and fourthly,

in the case of most isolated markers, including intra-cardiac echogenic focus, echogenic bowel, mild hydronephrosis and short femur, there is only a small effect on modifying the pre-test odds. Further studies are needed to establish reference ranges for each biometric marker and to estimate the effect of gestational age on screening performance. *(Agathokleous et al.)*

Second Trimester Down's and NTD Screening Reporting, Interpretation

The test report should be comprehensive, easy to understand; the risk prediction based on maternal age as well as biochemical screen should be clearly mentioned. Trisomy 21, and trisomy 18 should be covered along with NTD. There should be an explanatory note giving clarification of the test result, accuracy and limitations of the screening test, what are the post-test options and about the expected follow-up. This will help the referring physician to offer post-test counseling and further management (ref to second trimester screening report proformas in the annexure).

Screen positive (risk factor 1>100 or more): Confirmatory/ diagnostic testing (karyotyping of amniotic fluid cells) should be recommended in all aneuploidy screen positive pregnancies. An anomaly scan at 18–20 weeks, fetal echo- cardiography and second trimester screen test should be offered to all, especially those refusing fetal tissue sampling. Option of NIFTY or micro-array should also be mentioned, if available.

Screen negative (risk factor <1:500): Post-test counseling should be offered along with recommendations for anomaly scan (also fetal echo) at 20 weeks of pregnancy, when available.

Borderline reports (risk factor 1:100 to 1:500): Counseling, look for any of the high risk factors, get a second trimester anomaly scan. Option of amniocentesis for karyotyping should be discussed.

It is not advisable to recommend a repeat aneuploidy screening test as it does not improve the detection rate but can add confusion instead. However, for pregnancies screen-positive for Down's syndrome and dated by LMP, checking the accuracy of the gestational age is appropriate.

Integrated Screening Tests

In an effort to improve the DR while keeping FPR quite low, integrated test was recommended by many workers. In the integrated test first trimester combined test is done at 10–14 weeks and a quadruple/triple test done at 16–18th week and only at the end of these, a single risk prediction is given.

Sequential Screening

It is more efficient to carry out screening in both the first and second trimesters, in sequence. There are three types of sequential policy. Non-disclosure sequential screening uses first trimester PAPP-A and NT together with second trimester AFP, uE3, free β-hCG or hCG and inhibin-A. Risks are not used clinically until all markers have been tested. It has the substantial disadvantage that there is no early diagnosis or reassurance. Also there may be practical difficulties and ethical concerns over non-disclosure.

Stepwise sequential screening, begins with first trimester PAPP-A, free β-hCG or hCG and NT. Those with low risk have second trimester AFP, uE3, free β-hCG or hCG and inhibin-A; their risk is estimated from all seven markers. It is important to use a high first trimester cut-off than with non-sequential screening, otherwise the overall false-positive rate will be too high. And it is essential to use all seven markers together when calculating the final risk. It is invalid to ignore the first trimester markers at this stage although many practitioners are doing so because they do not have access to the appropriate risk calculation software. This policy restores some first trimester diagnosis.

Contingent sequential screening begins with first trimester PAPP-A, free β-hCG or hCG and NT. Women with very high

risk are offered immediate invasive prenatal diagnosis and only those with borderline risks are offered second trimester AFP, uE3, free β-hCG or hCG and inhibin; their risk is estimated from all seven markers. The borderline is chosen so that a large proportion of women have early assurance. This group has such a low risk that it very unlikely that further markers will lead to a final high risk result. In one illustrative analysis, the first trimester very high cut-off for stepwise and contingent is set to obtain an early detection rate of 70%, and the very low cut-off for contingent screening is set so that 85% of women have early assurance.

A non-disclosure policy had a detection rate 6–10% and 22–35% higher than the best first and second trimester serum combinations respectively. Stepwise screening was more efficient, yielding a slightly higher detection rate than non-disclosure screening whilst allowing considerable first trimester detection with a low first trimester false-positive rate. A contingent screening policy was the most efficient with a detection rate advantage similar to stepwise screening but required only about 15% of women needing second trimester tests. When hCG is substituted for free β-hCG in the second trimester of sequential screening the rates were not substantially altered. *Cuckle HS, Arbuzova S.*[8]

REFERENCES

1. Milunsky A, Canick JA. Maternal serum screening for neural tube and other defects. Genetic disorders and the foetus, Ed. Milunsky; John Hopkins university press; 2004; pp. 719–94.

2. Canick JA, Knight GJ, Palomaki GE. Low second trimester maternal serum unconjugated oestriol in pregnancies with Down's syndrome. Br J Obstet Gynaecol 1988; 95:330.

3. Bogart HM, Pandian MR, Jones OW. Abnormal maternal serum chorionic gonadotropin levels in pregnancies with foetal chromosomal anomalies. Prenat Diagn 1987; 7:623.

4. Van Lith JM, Pratt JJ, Beekhuis JR. Second trimester immunorective inhibin as a marker for foetal Down's syndrome. Prenat Diagn 1992; 12:801.

5. Cuckle H. Maternal serum screening for aneuploidy and Neural Tube Defects. Preventive Genetics, Gogate SG (Ed). Jaypee Brothers 2006; pp. 101–16.

6. Morris JK, Wald NJ. Quantifying the decline in the birth prevalence of neural tube defects in England and Wales. J Med Screening 1999; pp. 6–182.

7. Knight GJ. Maternal serum alpha protein screening techniques in diagnostic human biochemical genetics, Hommes FA, Ed. Weily-Liss, 1991:491.

8. Cuckle HS, Arbuzova S. Multi-marker maternal serum screening for chromosomal abnormalities. Genetic disorders and the foetus, Ed Milunsky A, John Hopkins University press 2004; pp. 795–835.

9. Agathakleos M, Chaveeva P, Poon LCY, Kosinski P, Nicolaides KH. Meta-analysis of second trimester markers for Tri 21. Ult Obstet Gynecol 2013; 41:247–261.

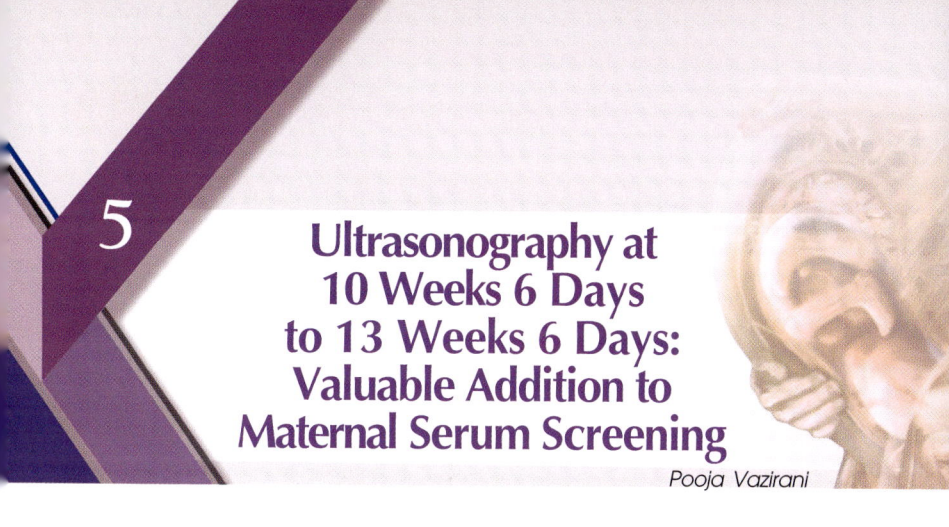

Pooja Vazirani

Ultrasonography at 10 Weeks 6 Days to 13 Weeks 6 Days: Valuable Addition to Maternal Serum Screening

The 11 – 13 + 6 weeks scan is a revolution in the field of fetal medicine. The main purpose of fetal imaging is to provide accurate information so as to facilitate best possible outcomes for mother and the fetus. Many chapters on this topic have been published in the past and so we do have enough literature about the 11 – 13 + 6 weeks scan.

However in this chapter we shall try to cover practical aspects of NT scan such as:

1. Aneuploidy screening
2. Structural scan-ISUOG guidelines
3. Screening for pre-eclampsia
4. 11 – 13 + 6 weeks scan for twins

ANEUPLOIDY SCREENING

10 weeks 6 days to 13 + 6 weeks: First trimester markers:

- Nuchal translucency (NT)
- Nasal bone (NB)
- Ductus venosus (DV)
- Tricuspid regurgitation (TR)
- Aberrant right subclavian artery (ARSA)
- FMF angle

Nuchal Translucency

Sonographic appearance of subcutaneous accumulation of fluid behind the fetal neck in the first trimester of pregnancy. The size of the nuchal translucency is more important than its appearance.

In trisomies 21, 18 and 13 the average NT in these defects is about 2.5 mm above the normal. In Turner syndrome, the median NT is about 8 mm above the normal median.

Increased nuchal translucency: Increased nuchal trans-lucency has been associated with chromosomal abnormalities and structural abnormalities.

Measurements of nuchal translucency:
- The fetal crown rump length should be 45–84 mm. The optimal
- Gestational age is 11 to 13.6 weeks
- The result from the transabdominal and transvaginal scanning are similar but reproducibility may be better with the transvaginal method
- A good sagittal section for the fetus as for measurements of fetal crown rump length should be obtained
- The magnification should be such that the fetus occupies at least three quarters (75%) of the image
- The nuchal translucency should be measured with the fetus in the neutral position
- The maximum thickness of the subcutaneous translucency between the skin and soft tissue overlying the cervical spine should be measured by the placing the calipers on the line as shown:

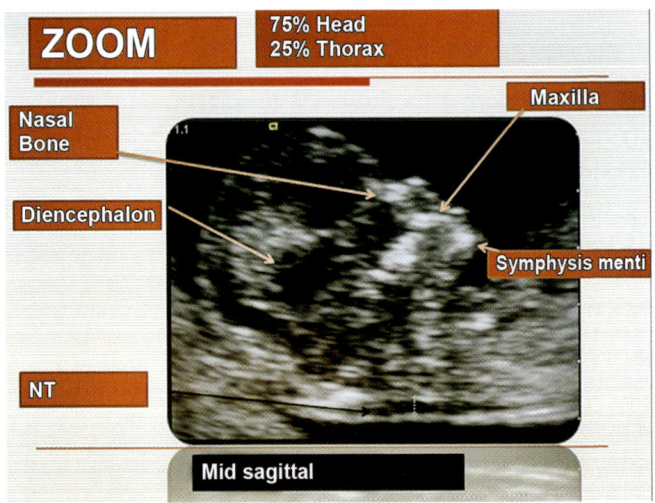

- Correct placement of calipers

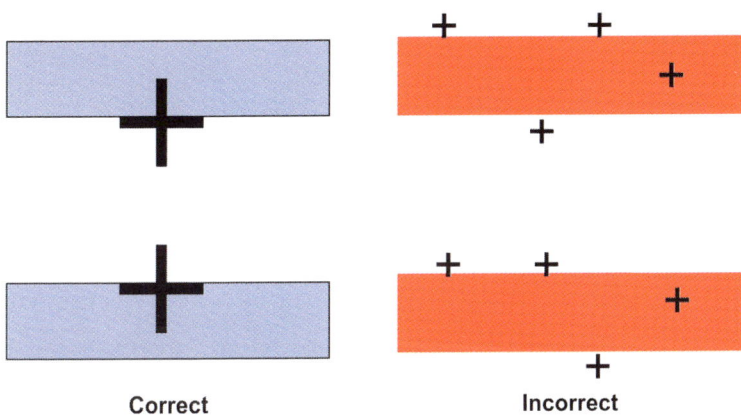

Correct **Incorrect**

- Care must be taken to distinguish between fetal skin and amnion because at this period of gestation both structures appear as thin membranes. This is achieved by waiting for spontaneous fetal movements away from the amniotic membrane; alternatively, the fetus is bounced off the amnion by asking the mother to cough and or by tapping the maternal abdomen.

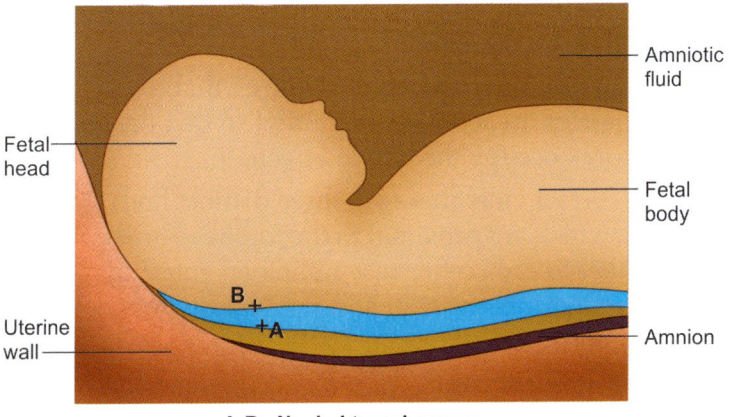

A-B= Nuchal translucency

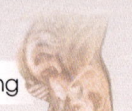

Increased nuchal translucency has been defined as NT above the 95th percentile for that particular CRL. However the value of 3.5 mm is at the 99th percentile irrespective of the CRL.

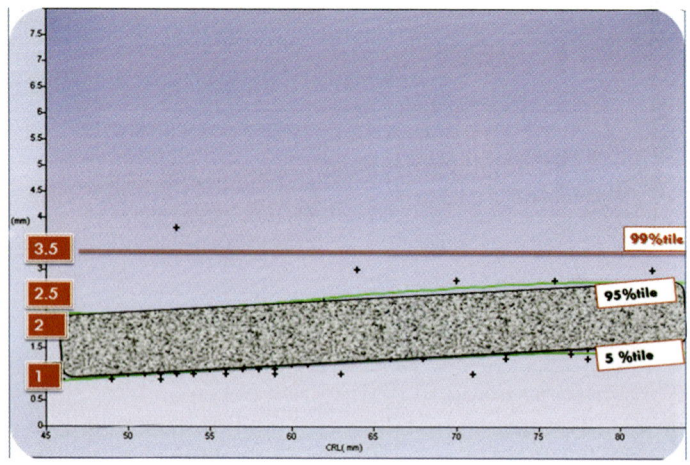

Nasal Bone

The nasal bone at the 11–1 4 weeks scan has recently received focus of attention as a marker for trisomy 21.60% of all T21 babies will have a depressed nasal bridge due to absence of the nasal bone. Thus a very important marker next to NT in the 11 – 13 + 6 weeks scan.

Examination of nasal bone requires that a midsagittal profile view of the fetus be magnified so that the head and upper thorax occupy the whole screen.

In the correct view there are three distinct lines. The first two lines which are proximal to the forehead are horizontal and parallel to each other resembling an equal sign (=) the top line represents the skin and the bottom one represents the nasal bone.

A third line almost in continuity with the skin but at a higher level represents the tip of the nose.

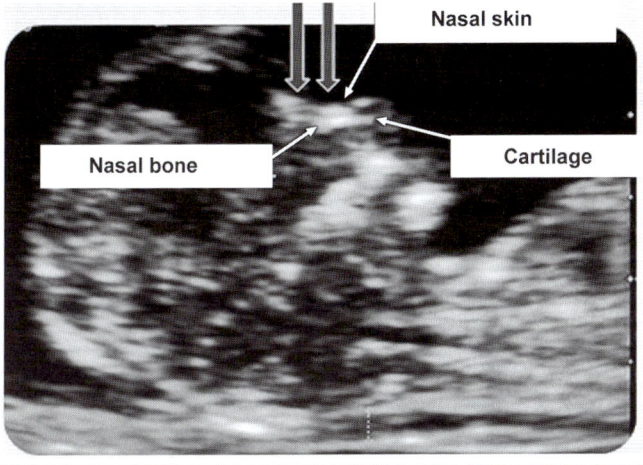

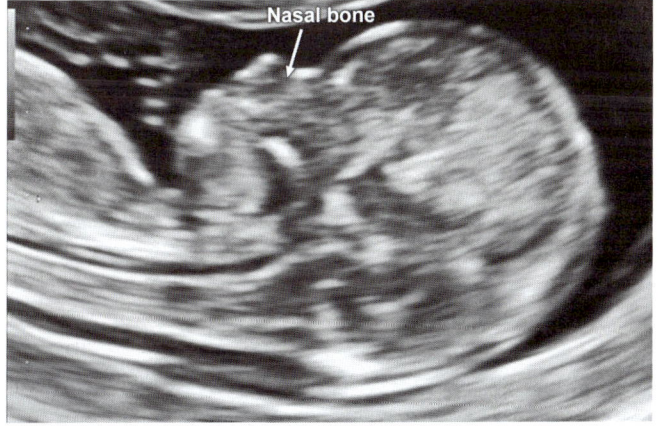

Nasal bone is absent in:

- 1.4% of the chromosomally normal fetuses
- 69% of fetuses with trisomy 21
- 50% of trisomy 18
- 30% of trisomy 13

In chromosomally normal fetuses the incidence of absent nasal bone:

- <1% in caucasian populations
- 10% in Afro-Caribbean's

Assessment of the nasal bone improves the detection rate from 90% to 93% with false +ve rate of 3–2.5%.

Ductus Venosus

- Midsagittal position of fetus
- Color box placed on the fetal trunk — then zoom
- Only trunk should be present in the zoomed image
- *PW Doppler*:
  - Gate of 0.5–1 mm only
  - Placed on the area of mosaic
  - Sweep speed — fast (4–5 wave forms)
  - Wall filter low — ideally 50–100

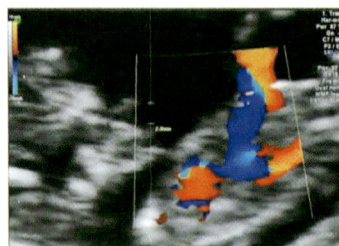
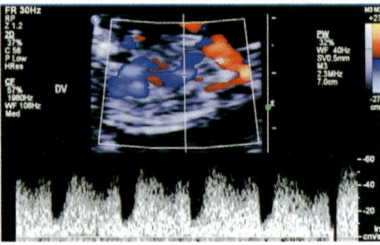
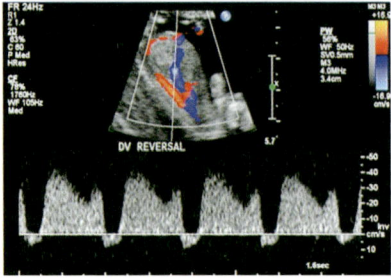

- If DV reversal is present in isolation then a combined FTS is advised and if low risk, then a 16 weeks scan for early TIFFA and fetal echocardiography is warranted.

Tricuspid Regurgitation

- Midsagittal position of fetus

- Only chest with 4 chamber heart should be present in the zoomed image
- PW Doppler:
  - Gate of 3 mm only on the tricuspid valve
  - Sweep speed — fast (4–5 wave forms)
  - Wall filter low — ideally 80–100

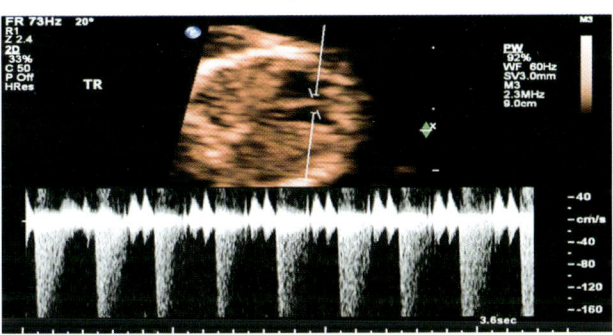

- If tricuspid regurgitation is present in isolation then a combined FTS is advised and if low risk, then a 16 weeks scan for early TIFFA and fetal echocardiography is warranted.

The above 4 marker are important and require training and certification process by the fetal medicine foundation. Amongst these the NT and the nasal bone hold significant importance with high LR value.

Method of screening	DR (%)
• Maternal age (MA)	30
• MA and serum biochemistry at 15–18 weeks	50–70
• MA and nuchal translucency (NT) at 11–13+6 weeks	70–80
• MA and NT and maternal serum free β-hCG and 13+6 weeks	85–90 PAPP-A at 11–
• MA and NT and nasal bone (NB) at 11–13+6 weeks	90
• MA and NT and NB and maternal serum free β-hCG and PAAP-A at 11–13+6 weeks	95

Increased NT and implications:

Definition of increased NT: Defined as measurement above the 95th percentile irrespective of whether the collection of fluid, septated or not and whether it is confined to the neck or envelopes the whole fetus.

Nuchal translucency	Chromosomal defect	Normal karyotypes	Major fetal anomalies	Alive and well
<95th percentile	0.2%	1.3%	1.6%	97%
95th–99th percentile	3.7%	1.3%	2.5%	93%
3.5–4.4 mm	21.1%	2.7%	10%	70%
4.5–5.4 mm	33.3%	3.4%	18.5%	50%
5.5–6.4 mm	50.5%	10.1%	24.2%	30%
>= 6.5 mm	64.5%	19%	46.2%	15%

Pathophysiology of increased NT:

- Cardiac dysfunction
- Venous congestion in the head and neck

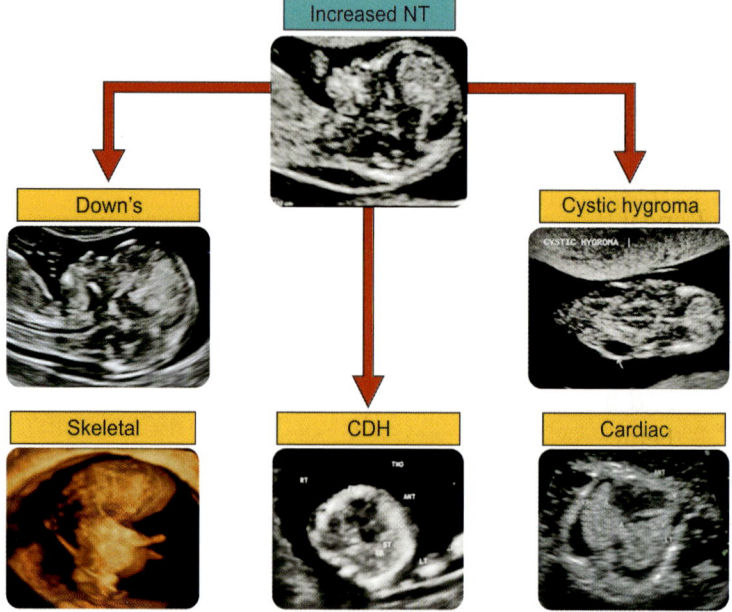

- Altered composition of the extracellular matrix
- Failure of lymphatic drainage
- Fetal anemia
- Fetal hypoproteinemia
- Fetal infection

Algorithm for stepwise management of increased NT:

1. Reconfirm the NT measurement
2. Check for other marker — nb, dv, tr
3. Classify NT → 95th percentile or >99th percentile. NT of 3.5 mm is at the 99th percentile for any CRL.

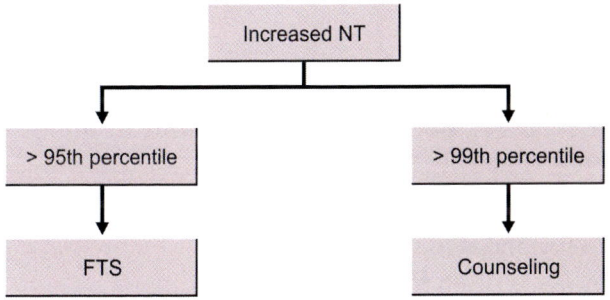

4. If NT >95th percentile and <99th percentile – combined FTS can be offered after counseling.

A. *If combined FTS is screen negative*:

- Recounsel and reassure
- Repeat the scan at 16 weeks with fetal echocardiography by an expert
- If normal – reassure for that period of gestation and repeat the scan at 20 weeks for TIFFA:
    - To rule out anomalies (2.5% risk)
    - 2nd trimester markers especially NF
    - Fetal echocardiography

B. *Direct test will be advised*:

1. If combined FTS is screen +ve

Or

2. Patient has agreed for direct test post-counseling

C. *Scenerio III*:
Combined FTS screen intermediate risk:

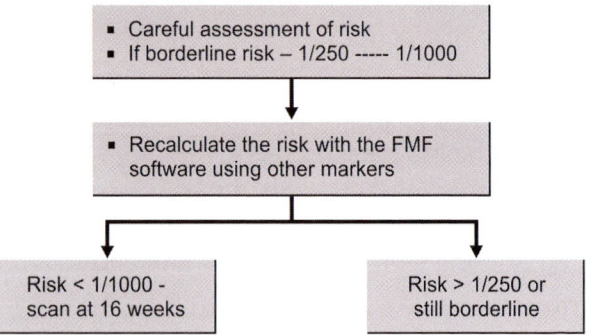

D. *Scenerio IV: NT >99th percentile*
- A fetal NT above 3.5 mm is found in about 1% of pregnancies

The risk of major chromosomal defects is very high:
- 20% for NT of 4.0 mm
- 33% for NT of 5.0 mm
- 50% for NT of 6.0 mm
- 65% for NT of 6.5 mm

Couples should be offered fetal karyotyping by CVS. If advanced in gestation then amniocentesis

The chance of an adverse pregnancy outcome such as miscarriage, fetal death and/or major fetal anomalies increases with increasing NT thickness. The prevalence of miscarriage or fetal death increases from 1.3% in the case of an NT between the 95th and 99th percentiles to about 20% for an NT of 6.5 mm or more. The prevalence of major fetal abnormalities is reported to increase exponentially, from 2.5% for an NT between the 95th and 99th percentiles to about 45% for an NT of 6.5 mm or more, major fetal abnormalities being defined as those requiring medical and/or surgical therapy. The chances of delivering a healthy baby are about

70% for an NT of 3.5–4.4 mm but about 15% for an NT of 6.5 mm or more.

Senat et al. found a developmental delay in 1.2% and an ASQ score <2 SD in 18% of the neonates with a history of increased NT, both of which were not significantly different from the results of the control group. In the white journal of 2007, *Bilardo et al.* 20 make an important contribution by reporting the follow-up of 451 euploid fetuses with a previously increased NT. The incidence of an adverse pregnancy outcome was related linearly to the initial degree of NT enlargement, ranging from 8 to 80%. A 20-week scan was performed in 425 cases. After a normal 20-week scan, an adverse pregnancy outcome was observed in 4% of the cases. Neurodevelopmental delay was diagnosed in 7/425 (1.6%) fetuses. *Bilardo et al.* conclude that, when findings at the 20-week scanare normal, a favorable outcome can be expected and parents can be reassured, regardless of the previous NT thickness. Their report of a low prevalence of developmental delay is in agreement with the study of *Senat et al.* 19 and several other studies.

In some cases of normal KT and increased NT at 12 weeks where a target scan is showing IUGR or persistent nuchal edema, then CGH array to rule out microdeletions and some syndromes is advised.

Assessment of Structure: Target Scan at 11–14 Weeks

The target scan is one of the most important scan during a pregnancy. It is done between 11 and 14 weeks.

The scan fulfils the following objectives:
1. It helps to predict the structural normalcy with confidence and reasonable limits of expectations.
2. Help to detect structural abnormalities-with expertise and good machine, up to 85–90% of structural abnormalities can be diagnosed.
3. To raise the suspicion of an abnormality that warrants serial scans.

The target scan compromises of 7 steps—stepwise approach to the fetus:

1. History
2. Survey
3. Biometry
4. Targetted imaging and aneuploidy screening
5. Fetal activity
6. Fetal environment
7. Reporting

Each step gives an information which finally is formatted into the report, that is the 7th step—reporting.

Stepwise Approach to the Fetus

1. History

It helps in identifying a high risk mother from a low risk mother. It helps in calculating the EDD in case their are irregular cycles. The background of the index child (previous child with problems) helps us in imaging the present fetus in a more relevant manner.

2. Survey

Its gives a global picture of the fetus. It helps is assessing the
a. Number of fetuses
b. Location of placenta
c. Space for the fetus, i.e. amniotic fluid.

3. Biometry

1. *CRL:* CRL can be done transabdominal and transvaginal. Midline sagittal section of entire fetus should be obtained. Ideally fetus should be horizontal in position. The image

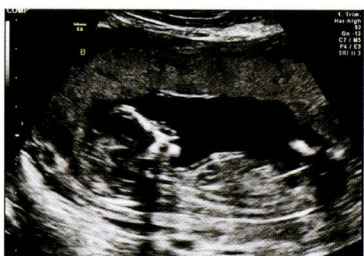
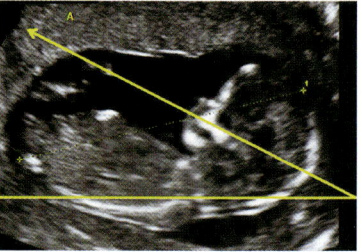

should be magnified to fill most of the width of the USG screen.

2. *BPD, HC, AC and FL:* Just like the 2nd trimester biometry.

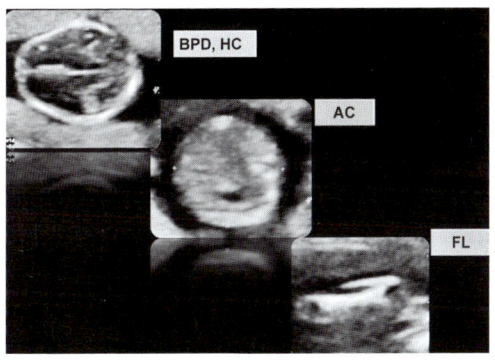

Fetal Anatomy

The early anomaly scan at 11–13+6 weeks has its own advantages

Early detection and exclusion of many major anomalies

Early reassurance to at risk mothers

Early genetic diagnosis

Easier pregnancy termination if appropriate.

Its limitations:

1. Trained and experienced personnel

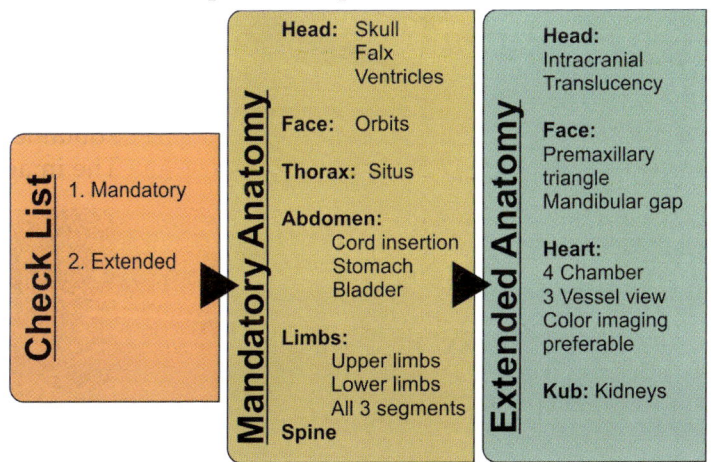

2. Late developments of certain anatomical structures and pathologies

The fetus has to be examined in 3 planes in each section of the head, face, trunk.

Head

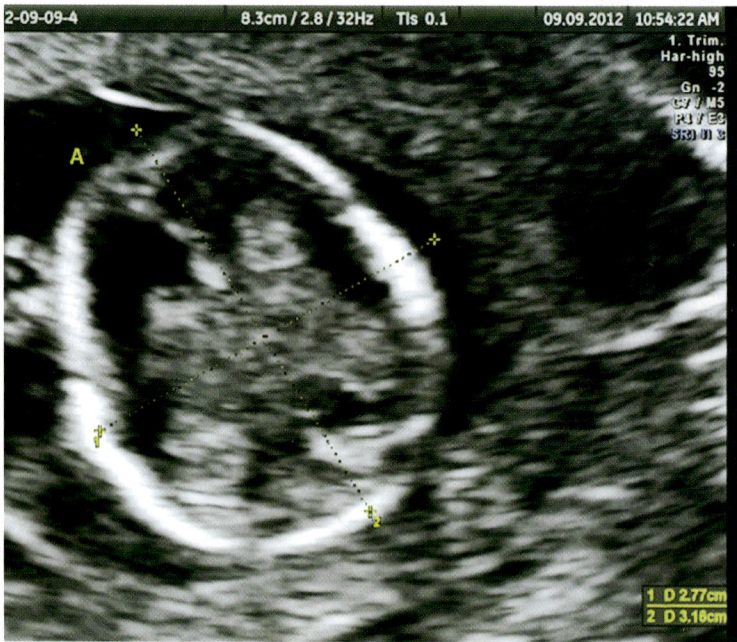

Three structures seen in this plane:

a. Intactness of skull bone

b. Midline falx

c. The lateral ventricles—Choroid plexus

Abnormalities of the head at 1st trimester:

1. Anencephaly/acrania

2. Encephalocele

3. Holoprosencephaly

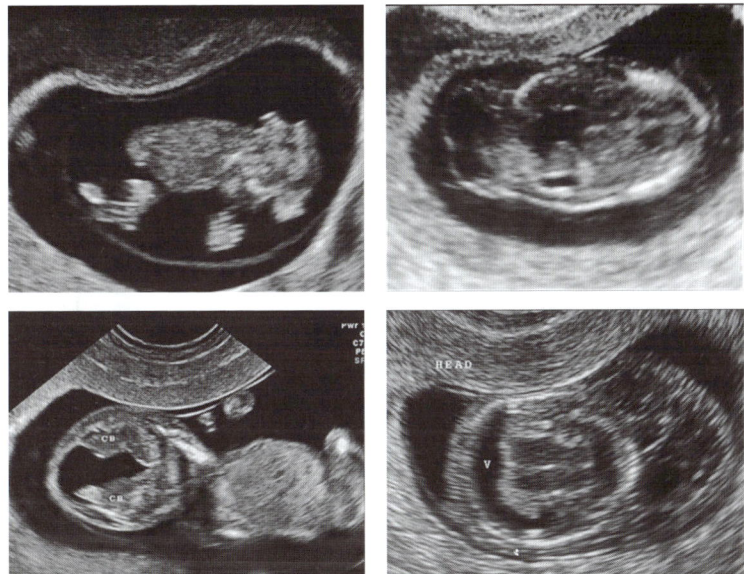

Intracranial translucency:

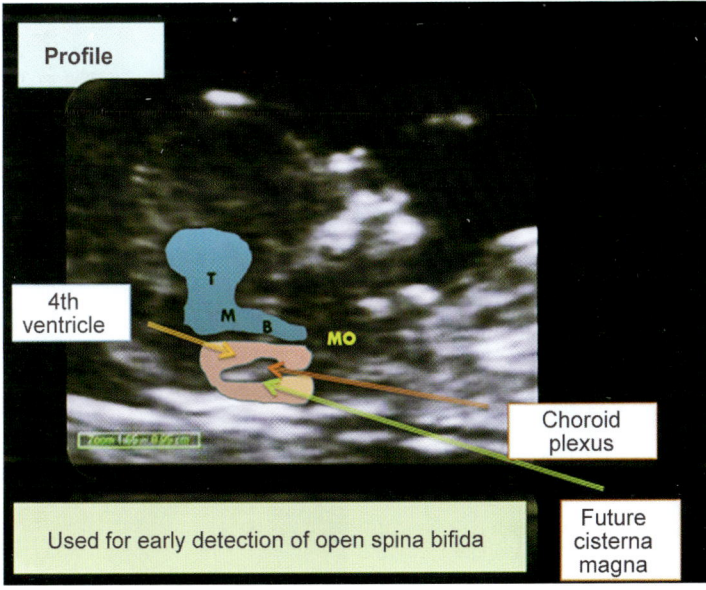

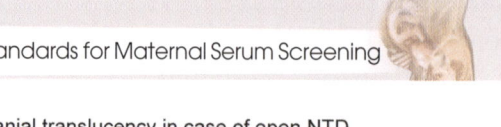

Abnormal intracranial translucency in case of open NTD

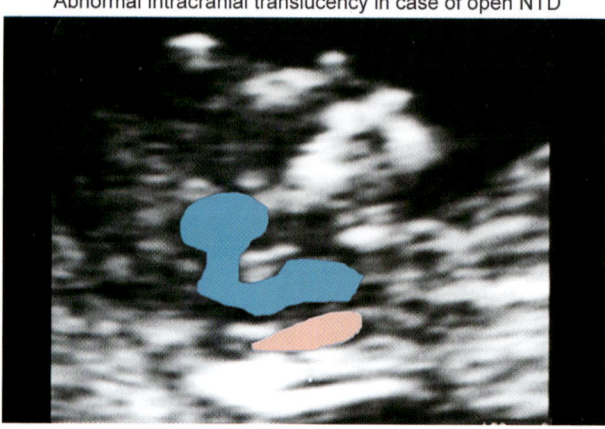

Examination of the midsagittal view of the fetal face is performed routinely for assessment of fetal NT and the nasal bone in screening for aneuploidies.

If in this same view the fourth ventricle is not visible—undertake detailed examination of the fetal spine due to the high possibility of an underlying open spina bifida.

3. *Face*:

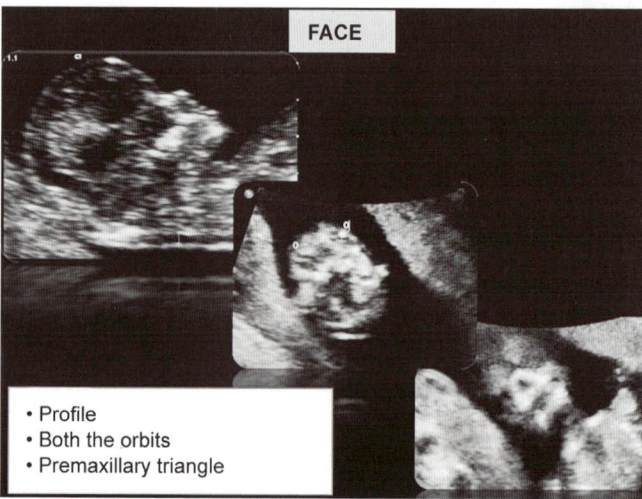

FACE

• Profile
• Both the orbits
• Premaxillary triangle

Three structures:
A. Midline—nasal bone, maxilla and tip of mandible
B. Orbits with lens (optional)
C. Premaxillary or the retronasal triangle—helps in ruling out CL CP

Abnormalities of the face:

1. Hypotelorism
2. Cleft lip and palate
3. Midline hypoplasia
4. Micrognathia

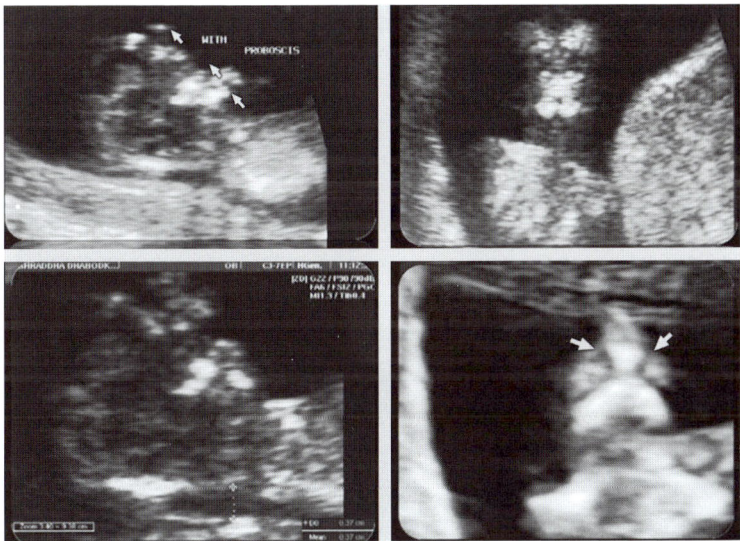

4. *Thorax:*

- Lungs are still developing
- Heart—though the development is complete, diagnosis of all cardiac problems may not be accurate at 11–14 weeks
- Major cardiac defects can be diagnosed

Cases where screening for cardiac abnormality is required are:

1. Increased NT
2. DV reversal

3. TR is +ve

Fetal echocardiography at 16 weeks is mandatory.

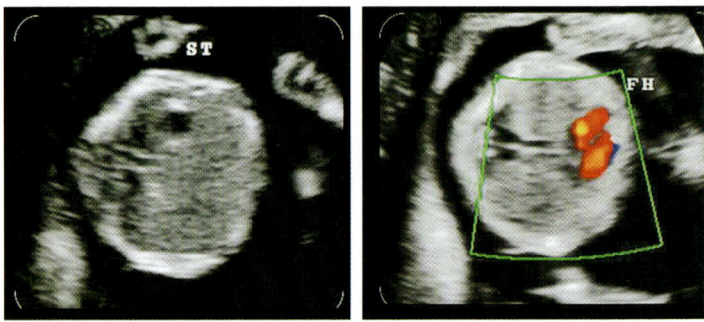

Abnormalities of the heart: CRL >55–60 mm
1. Univentricular heart
2. Single outflow
3. Malaligned VSD
4. AVSD
5. Tetralogy of Fallot
6. Tricuspid atresia, etc.

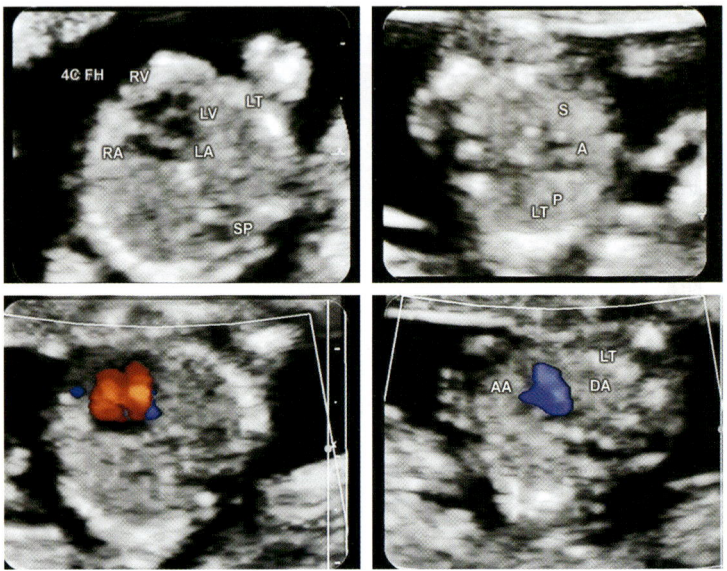

Heart:
1. Situs
2. If 4 chambers and outflow tracts with color—important
3. If 4 chambers and outflow tracts with b mode possible then can be attempted

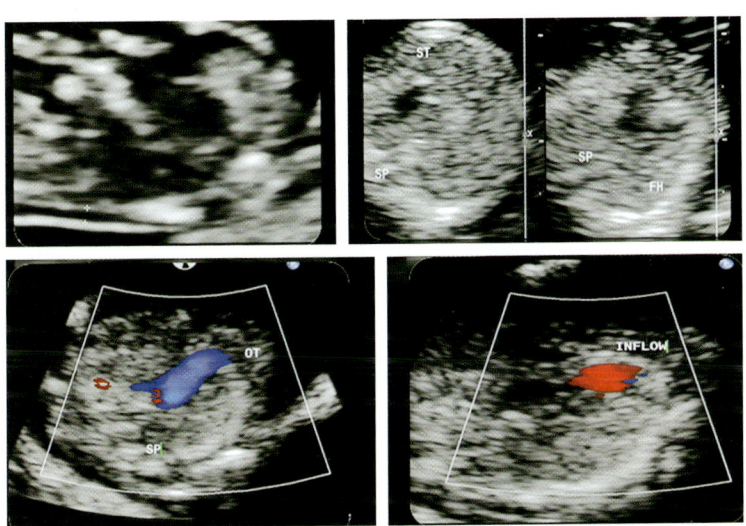

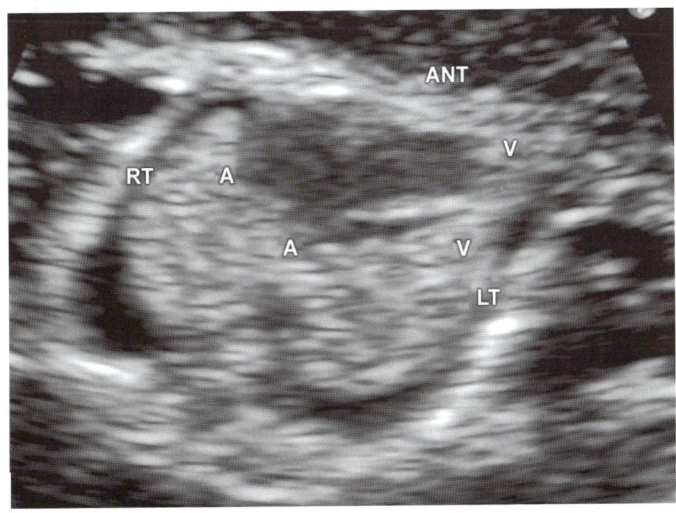

Abdomen

- Presence of stomach important than absence
- Physiological omphalocele closes by CRL of 44 mm
- Pathological omphalocele—those beyond CRL of 45 mm or containing liver from the onset irrespective of CRL

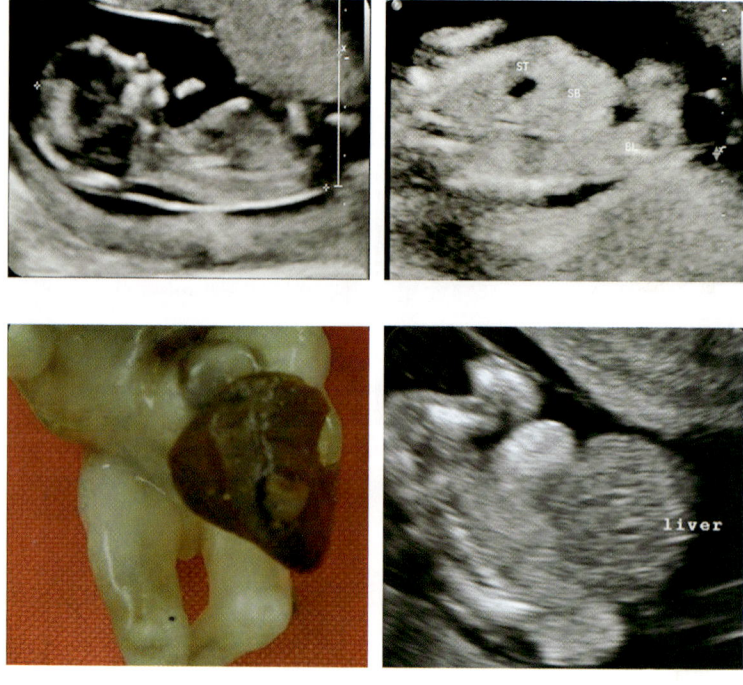

5. *Genitourinary system:*

- Bladder is seen by 11 weeks
- Normal size is up to 6 mm
- Megacystis — beyond 7 mm
- 7–15 mm — high risk for chromosomal
- >15 mm — high risk for obstructive pathology

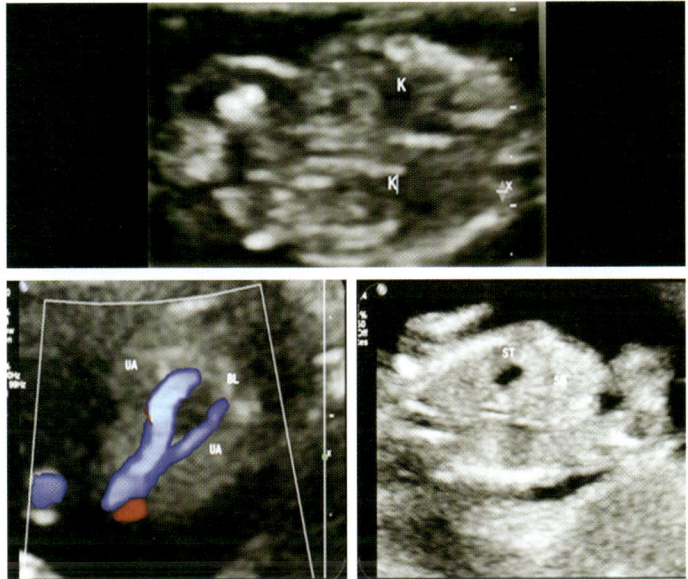

Abnormalities of genitourinary system: Megacystis— urethral atresia or posterior urethral valves.

Prune Belly syndrome:

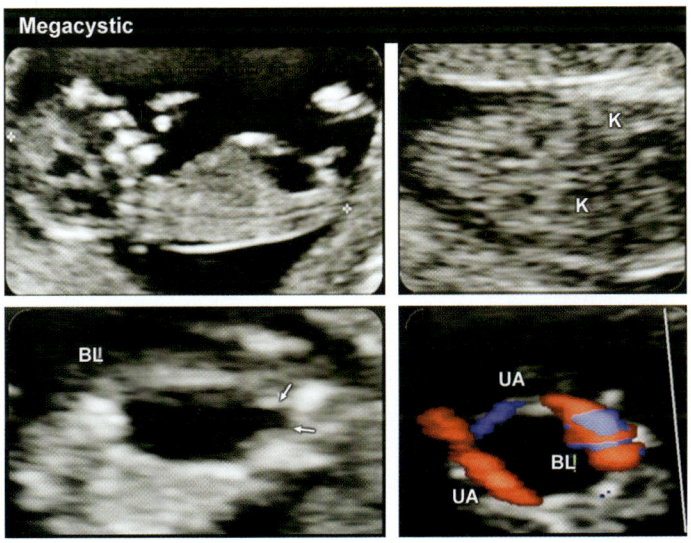

Prune Belly syndrome

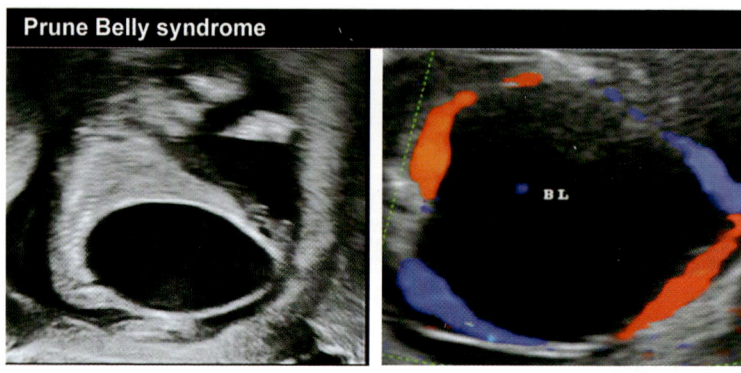

6. *Skeletal system*:
- *Spine*:
    - Coronal/Sagittal view
    - Indirect sign to rule out open spinal defect—intracranial translucency
- *Extremities*:
    - Upper and lower limbs
    - Three segments—Proximal, midsegment, distal

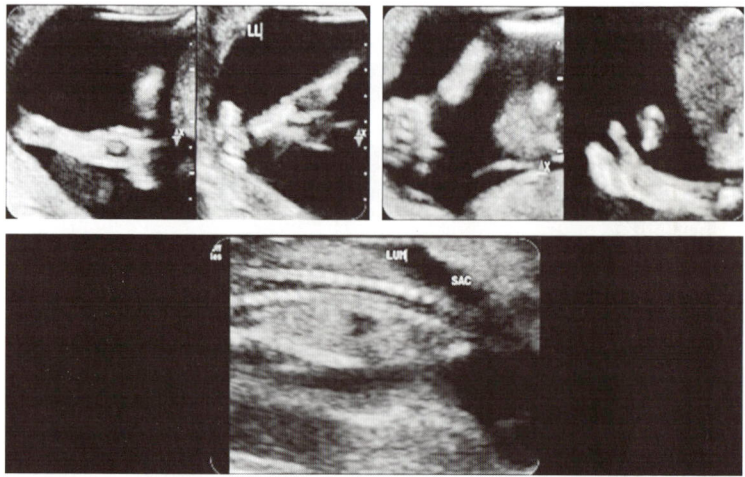

- *Skeletal defects*:
    - All bones are formed by this time

- Absence of any segment can be diagnosed—look for all segments
- Lethal skeletal dysplasias with deformed bones can be diagnosed
- In a case with previous h/o skeletal problem—diagnosis at 11–14 weeks possible

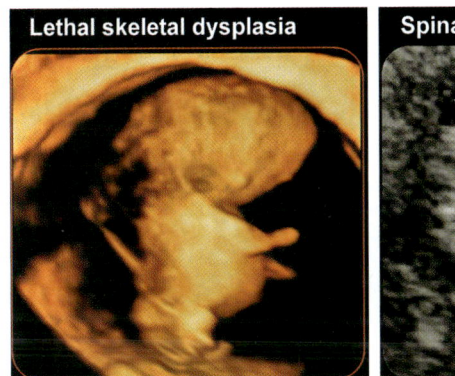

Lethal skeletal dysplasia

Spina bifida

Spine

Syndromes in First Trimester

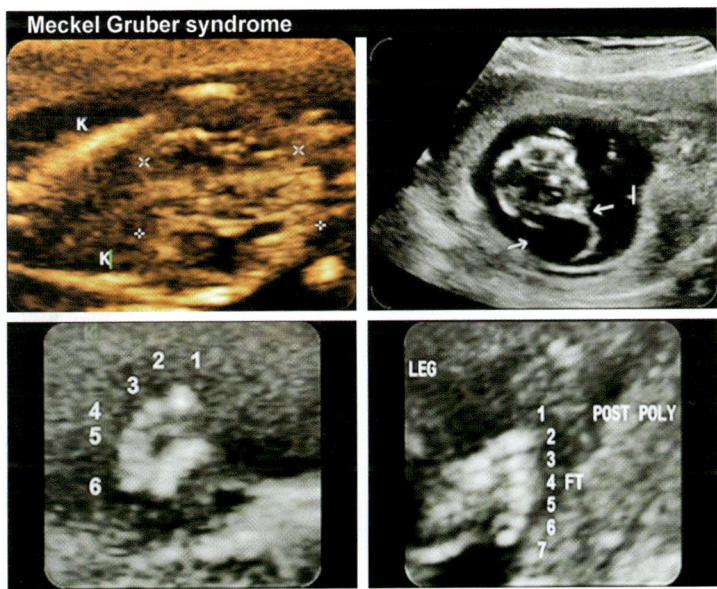

Meckel Gruber syndrome

LEG

POST POLY

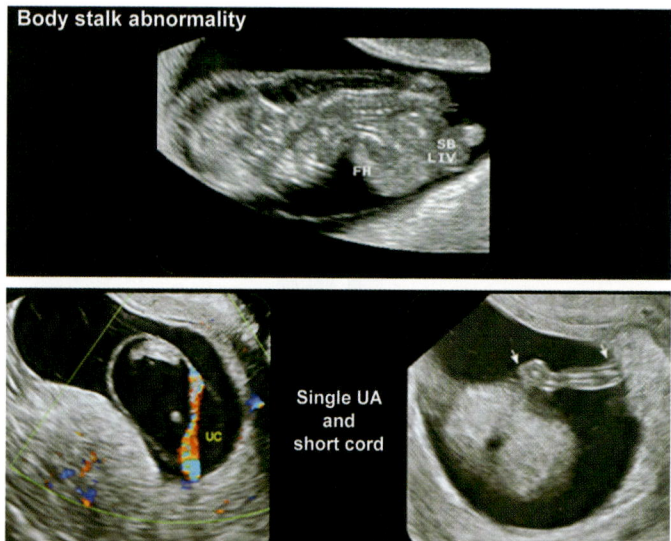

Imaging in twins: It is the best time to evaluate twin gestation in terms of the chorionicity and amnionicity.

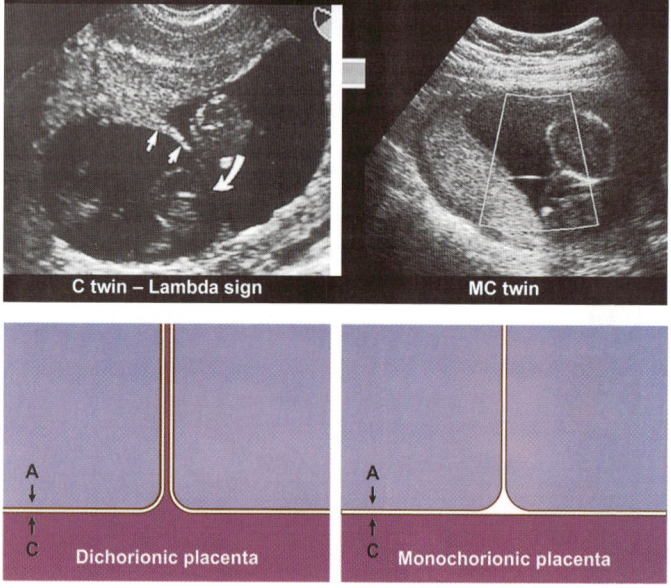

7. *Screening for pre-eclampsia*:
- Incidence of pre-eclampsia is 3–4% worldwide
- Accounts for 10,000 maternal deaths worldwide
- 2nd most common 4 fold increase in perinatal mortality
- 10%—stillbirths, 15%—preterm births
- Causes long term maternal and infant morbidity and cause of maternal death in developing countries.

Many pregnancy complications can now be predicted at an integrated first hospital visit at 11–13 weeks

The traditional pyramid of care should be inverted with the main emphasis placed in the first rather than third trimester of pregnancy, hypertension.

—Prof Nicoloidas

Prediction if Pre-eclampsia

Uterine artery Doppler:

Lowest uterine artery PI is taken

Sample gate—2 mm

Angle—30°

3 consecutive similar waves forms are taken

Maternal mean arterial blood pressure—3 measurements are taken and mean is taken

Biochemistry—PAPPA and PLGF.

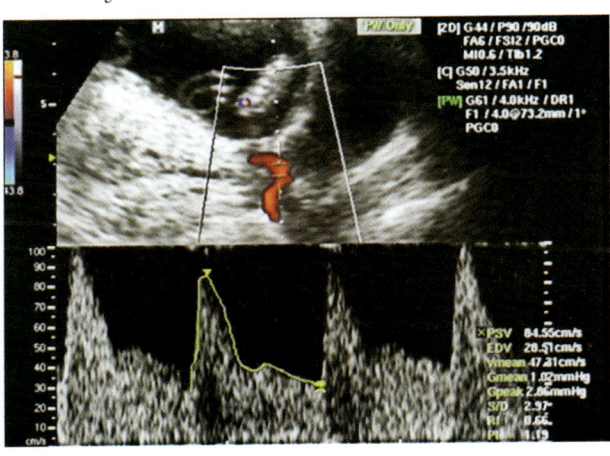

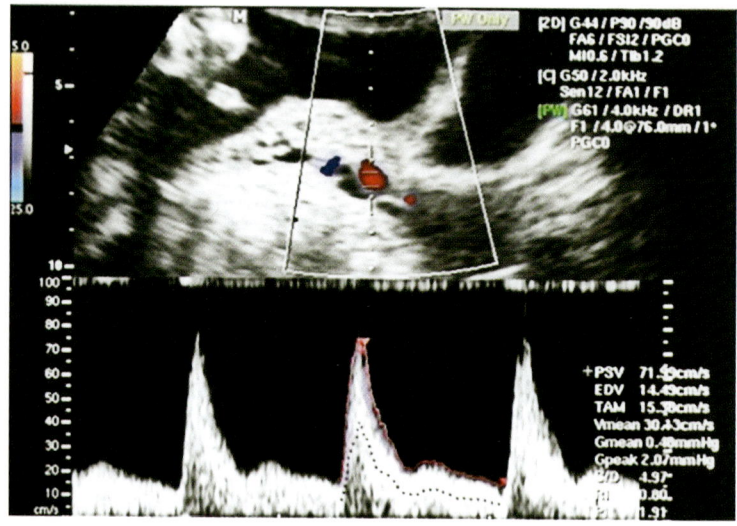

Pre-eclampsia

Prevalence:
- <34 weeks –1/200
- <32 weeks –1/300

	severe PET	very severe PET
14 weeks	40%	60%
23 weeks	77%	93%

- At 14 weeks efficacy equal to DS screening
- At 23 weeks > DS screening

Risk Prediction

Screening test	Risk prediction for early PE
• Maternal factors	• 33.0%
• Maternal factor plus PAPP-A	• 47.0%
• Uterine artery PI	• 54.1%
• MAP	• 49.7%
• Maternal factors plus biophysical markers plus PAPP-A and PLGF	• 77.8%

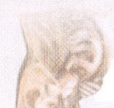
Best predictor for pre-eclampsia:

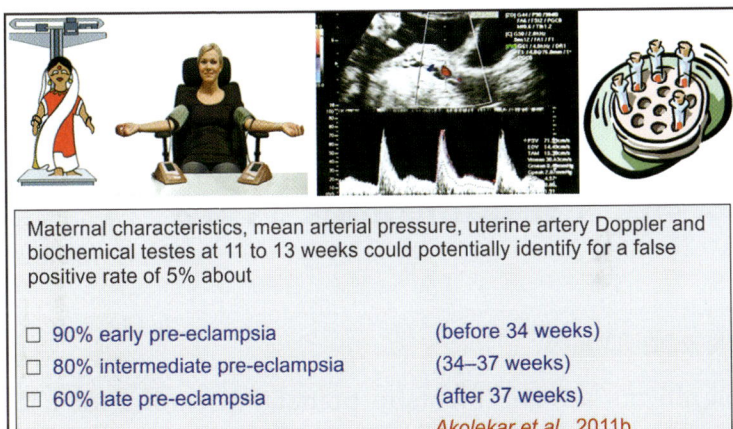

Maternal characteristics, mean arterial pressure, uterine artery Doppler and biochemical testes at 11 to 13 weeks could potentially identify for a false positive rate of 5% about

☐ 90% early pre-eclampsia   (before 34 weeks)

☐ 80% intermediate pre-eclampsia   (34–37 weeks)

☐ 60% late pre-eclampsia   (after 37 weeks)

Akolekar et al., 2011b

REFERENCES

1. Fetal medicine foundation 11 – 13 + 6 weeks scan.

2. Increased NT, normal Kt, *Ultrasound Obstet Gynecol,* 2007.

3. Increased NT and what next *Ultrasound Obstet Gynecol,* 2009.

4. Increased NT what should we be telling patient, *Prenatal Diagnosis* 2010.

6

Non-Invasive Prenatal Testing (NIPT) as an Advanced Screening for Aneuploidies of Chromosomes 13, 18, 21

Gayatri Jayaraman

INTRODUCTION

Significant advancements in prenatal screening for fetal aneuploidy have been made over the past few decades. Initially, advanced maternal age was the indication for which pregnancies were assessed for risk of fetal chromosomal abnormality. Fewer than 30% of trisomy 21 pregnancies were detected and of the screen positive women opting for invasive prenatal diagnosis, only 2% had karyotype abnormalities.[1] The low diagnostic yield was comparable to the procedure related fetal loss of 0.5%–1% associated with amniocentesis and chorionic villus sampling (CVS).[2] With the introduction of maternal serum markers, the detection rate more than doubled and the diagnostic yield from invasive testing increased to about 4%.[3] Combining ultrasound markers like nuchal translucency (NT) thickness along with first trimester serum markers has further increased the detection rate to approximately 90%. However, only about 5% of screen positive women carry a fetus with trisomy 21 indicating that the majority of invasive tests are unnecessary.[4, 5] Hence, there has been a constant desire to develop a screening test with better performance especially, lower false positives.

Discovery of fetal cells in maternal blood provided an exciting opportunity for non-invasive analysis of fetal DNA. Further studies revealed very low amounts of circulating

fetal cells among maternal cells in the order of 1 in 105 to 109.[6] This approach has been largely unsuccessful to reliably detect and isolate fetal cells.[7,8] In 1997, cell free fetal DNA was discovered in maternal plasma which was more abundant than fetal cells. Though designated as 'fetal', the cell free fetal DNA is placental in origin and results from apoptosis of trophoblast cells.[9] Cell free fetal DNA can be detected as early as four weeks of gestation.[10] Studies initially suggested that fetal cell free DNA constitutes only 3 to 6% of total cell free DNA in maternal circulation.[11] However, more recent studies reveal that the 'fetal fraction' may be closer to 10–20%.[12] Cell free fetal DNA does not persist, unlike fetal cells, and is not detectable soon after birth. This obviates possible confounding by previous pregnancy in very close successive pregnancies.[13] Subsequent studies and clinical trials led to the commercial launch of non-invasive prenatal testing (NIPT) in 2011 and soon after, the second generation NIPT in 2012.

Technical Methodologies of Non-invasive Prenatal Testing

Currently, two different methods are commercially available- a quantitative method and a single nucleotide polymorphism (SNP) method. The quantitative method is commercially available through sequenom™, Beijing Genomics Institute (BGI)™, Verinata™ and Ariosa™. Quantitative NIPT uses either massively parallel shotgun sequencing (MPSS) or targeted sequencing. The SNP method is commercially available through Natera™. This is the second generation NIPT that uses targeted sequencing of polymorphic regions on chromosomes of interest.

Quantitative Methodology

Cell free DNA fragments are sequenced and mapped to a specific location on the human genome and these mapped sequences are referred to as 'tags'. The number of tags from a chromosome of interest is compared to those from a reference chromosome that is assumed to be disomic. This ratio is compared to the ratio expected for a pregnancy with a euploid fetus. If fetal aneuploidy is present, there would be a relative excess or deficit of the chromosome of interest compared to the reference chromosome which would alter the expected normal ratio. It is important to note that the chromosome counts derive from both maternal and fetal DNA and cannot be treated differently by the statistic algorithm. This implies that any deviation noticed could either be due to fetal contribution or maternal contribution. Since, this is missed by the statistical algorithm, there could be false positive result for a chromosomally abnormal mother who has a euploid fetus.[14,15]

There are two approaches to quantitative NIPT: Massively parallel shotgun sequencing (MPSS) and targeted sequencing.

a. *Massively parallel shotgun sequencing*: The first 36 nucleotides or base pairs (bp) of millions of fetal and maternal cell free DNA fragments are sequenced. This sequence of 36 bp is usually unique enough and can be mapped to a specific location on the human genome. Approximately, 12 to 25 million tags are available per sample. An excess or deficit in the number of counts from the chromosome of interest as compared to the reference chromosome is expressed as normalized chromosome values (NCV) or z-scores. Aneuploidy is detected and called for the chromosome if NCV > 4.0 or if z-score >2.5.[15] Companies offering MPSS-based quantitative NIPT include sequenom™, Beijing Genomics Institute™ and Verinata™. These tests are called as Materni T21™, NIFTY™ and verifi™ respectively.

b. *Targeted sequencing*: This method selectively sequences the DNA fragments only from the chromosomes of clinical importance — chromosomes 13, 18 and 21. This approach requires considerably lower sequencing and thereby reduces the cost as well. In this method, 192 polymorphic loci are amplified for DNA fragments mapped to chromosomes 1–12. The loci where the fetal allele differs

from the maternal allele are used to estimate the fetal fraction. DNA fragments from chromosomes 13, 18, 21, X and Y are selectively sequenced and counted. The risk algorithm includes fetal fraction, sequencing results and maternal age to provide patient-specific risk for aneuploidy.[14,15] Ariosa™ offers the targeted sequencing based quantitative NIPT and the test is called as Harmony™.

Single Nucleotide Polymorphism-based Methodology

A single nucleotide polymorphism (SNP) is one of the most common genetic variations wherein a single nucleotide in a DNA sequence is different among individuals. For example, one individual would have a Cytosine (C) nucleotide on a specified DNA location while another would have a Guanine (G) instead. The two versions are called as alleles and the combination of the alleles on the two chromosomes is called as the genotype. Fig. 6.1 illustrates the above.

The SNP-based NIPT method utilizes 19,488 single nucleotide polymorphisms (SNPs) on chromosomes 13, 18, 21, X and Y. This method uniquely allows for distinction between maternal and fetal cell free DNA in maternal plasma. Maternal blood is centrifuged and cell free DNA is isolated from maternal plasma and buffy coat. Maternal plasma contains cell free DNA of maternal and fetal origin while the buffy coat contains only maternal cell free DNA.

	SNP ↓		
Individual 1	A C T C A C T A C T G A C T	Paternal chromosome 21 Maternal chromosome 21	Genotype CG
Individual 2	A C T C A C T A C T C A C T	Paternal chromosome 21 Maternal chromosome 21	Genotype CC
Individual 3	A C T G A C T A C T G A C T	Paternal chromosome 21 Maternal chromosome 21	Genotype GG

Fig. 6.1: SNP and genotype

An optional sample of paternal cheek swab can be collected to isolate paternal DNA. Cell free DNA isolated from these samples are sequenced to collect genotypic information on 19,448 SNPs. The proprietary NATUS (next-generation aneuploidy test using SNPs) algorithm is designed to infer the fetal genotype after 'subtracting' maternal genotype. The algorithm incorporates maternal genotype and recombination frequencies to construct billions of theoretical fetal genotypes. After considering the position of the SNPs and the possibility for recombination, a maximum likelihood is calculated for the possible scenarios—fetus is normal, aneuploid (trisomy 13, trisomy 18, trisomy 21 or sex chromosome) or triploid. Personalized risk scores are provided for these conditions.[14, 15]

The different NIPT technical methodologies have been depicted in Flowchart 6.1.

Flowchart 6.1: Different technical methodologies to sequencing cell free DNA in plasma of pregnant women.[15]

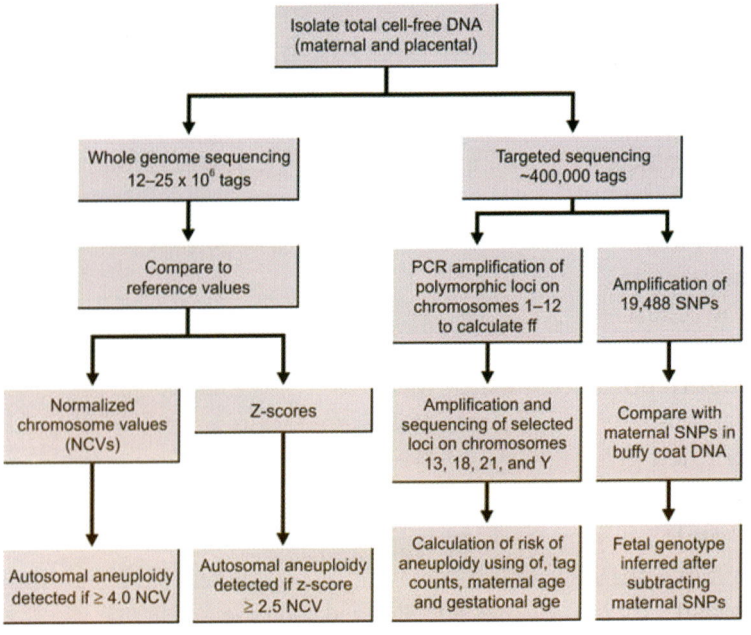

Factors Affecting Accuracy of Non-invasive Prenatal Testing

Fetal Fraction

Maternal plasma contains both maternal and fetal cell free DNA. Fetal fraction is the proportion of fetal cell free DNA amongst total cell free DNA in maternal plasma.

$$\text{Fetal fraction} = \frac{\text{Fetal cell free DNA}}{\text{Maternal cell free DNA} + \text{fetal cell free DNA}}$$

Fetal fraction is a critical factor that influences the sensitivity of NIPT. The higher the fetal fraction, the easier it is to detect aneuploidy. Low fetal fractions (<8%) are associated with lower sensitivity in quantitative method. Studies show that the sensitivity for Trisomy 21 can be significantly reduced to only 75% when the fetal fraction is below 8% in the quantitative method.[16] Fetal fraction can also influence the specificity of NIPT for quantitative NIPT as it impacts the number of counts for the chromosome of interest in an aneuploid fetus. High fetal fraction would generate higher counts resulting in increase in detection rate but mostly no impact on the false positive rate. However, with the SNP method, increase in fetal fraction will be associated with increase in detection rate and wouldmost likely reduce the false positive rate as well.This difference is attributed to the kind of data that is generated by the quantitative method as compared to the SNP method. In the quantitative method, the cell free DNA contributions made by the mother and the fetus are assumed to be equal. However, in the SNP method, highly polymorphic SNPs on chromosomes of interest are targeted. Higher fetal fractions would result in more clearly defined genotypic patterns in both euploid and aneuploid fetuses, hence increasing detection rate and reducing false positive rate.[17]

Several factors influence fetal fraction and hence the sensitivity of NIPT.

Factors Affecting Fetal Fraction

a. *Maternal weight*: Maternal weight is the most significant demographic factor that affects fetal fraction. Maternal weight is inversely proportional to fetal fraction with an increase in maternal weight resulting in decrease in fetal fraction. Obese pregnant women have increased maternal cell free DNA as a result of remodeling of adipose tissue. Though, the amount of fetal cell free DNA is unaffected, the increased maternal cell free DNA results in an overall reduction in fetal fraction.[15] A study on pregnant women at 11–13 weeks gestation showed than women who weighed 60 kgs had a median fetal fraction of 11.7% which decreased to 3.9% in women who weighed 160 kgs.[18]

b. *Gestational age*: Fetal fraction increases with gestational age with an increase of 0.1% per week between 10 to 21 weeks gestation and thereafter increases by 1% per week. In a study of more than 22,000 pregnancies, 1.9% of samples needed to be redrawn because fetal fraction was too low for accurate analysis. In 56% of pregnancies, where a redrawn sample was obtained, the fetal fraction was above the required 4%. Maternal weight and gestational age accounted for approximately 27% of the variation in fetal fraction.[19]

c. *Mosaicism*: Confined placental mosaicism (CPM) is relevant to NIPT because the cell free 'fetal' DNA originates from apoptosis of cytotrophoblasts. CPM can result in false positive or false negative results in NIPT. NIPT could fail to detect a viable trisomic fetus with a substantially low percentage of abnormal cytotropho-blasts. Similarly, the cytotrophoblasts may be mostly abnormal with a normal fetus, leading to false positive NIPT results.[15] Detection of CPM may be possible when the fetal fraction is high. In the MELISSA trial, mosaic cases for trisomy 18 and trisomy 21 with at least 29% were detected.[20] Maternal mosaicism could also result in false positive results in quantitative methods as distinction

between maternal and fetal cell free DNA is not made. A study revealed that 8.56% of false positives for sex chromosome abnormalities using quantitative methods were attributed to maternal mosaicism.[21] Maternal mosaicism could also potentially be detected with SNP-based NIPT since distinction between maternal and fetal cell free DNA is possible through SNP genotyping.[15]

d. *Fetal aneuploidy*: Pregnancies with trisomy 21 fetuses have higher fetal fraction than euploid fetuses. However, fetuses with trisomy 13, trisomy 18 and monosomy X have decreased fetal fraction. This might explain the higher sensitivity of the quantitative method for trisomy 21 as compared to trisomy 13 and trisomy 18.[15]

Multiple Gestations

There have been studies evaluating the performance of NIPT in twin pregnancies which have used the quantitative method. However, quantitative methods do not differentiate between cell free DNA contributions made by individual fetuses resulting in false positives for discordant twin pregnancies. One study included 5 twin pregnancies wherein NIPT called 2 cases positive for Down syndrome and 3 cases negative for Down syndrome. Positive NIPT cases were followed up with invasive testing which confirmed correct calls for both positive cases. However, karyotyping revealed that one of the cases was a discordant twin pregnancy for Down syndrome, resulting in a false positive for the normal twin.[22] In another study on 12 twin pregnancies, 11 were normal pregnancies and only one discordant twin pregnancy for Down syndrome. All pregnancies were correctly called except for a false positive for the normal twin in the discordant twin pregnancy case.[23]

In twin pregnancies, discordant for trisomy, the excess cell free DNA fragments from the aneuploid fetus may mask the normal cell free DNA fragments from the normal fetus, leading to false positive results for the existing fetus.

Additionally, in a singleton pregnancy that had a vanishing twin, there is a possibility that cell free DNA from the resorbed twin interfered with result for the living twin.[14] This has to considered as more than 15% of discordant results using the quantitative method involved vanishing twins.[24] This problem might be overcome by using the SNP-based method. Using SNPs can determine zygosity and examine distribution of fetal fragments in dizygotic twins and also detect vanishing twins.

NIPT for multiple gestations is currently being offered commercially by quantitative method and under validation for the SNP-based method.

Triploidy

Triploidy can be detected on ultrasound through characteristic sonographic features like multiple congenital anomalies, intrauterine growth restriction, increased nuchal translucency thickness and oligohydramnios. Though most cases of triploidy undergo early spontaneous abortion, the incidence of triploidy at 10–14 weeks is about 1 in 3000.[25]

Triploid pregnancies can arise as a result of an additional set of chromosomes from the mother (digynic) or from the father (diandric). The quantitative method might fail to detect triploidy, especially digynic triploidy, as there is a proportionate increase in all chromosomes and the ratios between chromosome of interest and the reference chromosome may remain undisturbed.[14] Also, in digynic triploid pregnancies, the placenta is very small and hence the fetal fraction might be insufficient. This has been reflected in a study where 3 out of 9 digynic triploid pregnancies (33%) had insufficient fetal fraction.[26] The SNP-based NIPT can detect triploidy and also differentiate between digynic and diandric triploidy with the use of SNP genotypes. This could potentially be useful in the future to assess the risk for choriocarcinoma and recurrence in future pregnancies associated with diandric triploidy and digynic triploidy respectively.[27,28]

Chromosome Biology

Human DNA comprises of four types of nucleotides—adenine, guanine, cytosine and thymine. GC content is the percentage of guanine and cytosine nucleotides in the total amount of nucelotides in a fragment of DNA or complete genome. The average GC content in the human genome is 41%.[29] GC rich and GC poor regions tend to have lower or no coverage of sequencing reads in MPSS platforms.[30] Chromosomes 18, 21 and X have a GC content in the mid range and are hence well covered with the quantitative method. However, chromosome 13 has one of the lowest GC contents and the sensitivity is lower with the quantitative method. For this reason, approaches have been used to normalize the GC ratios for chromosome 13 to increase the sensitivity.[15] The sensitivity and specificity for the common trisomies by the different NIPT tests as evaluated in the clinical validation studies are compared in Table 6.1. The positive and negative predictive values of the tests would depend on factors like disease prevalence which is directly related to maternal age.

The performance and coverage of the quantitative method and the SNP-based method are summarized in Table 6.2.

Advantages and Limitations of NIPT

Advantages

- Safe test with no risk for pregnancy loss
- Offers higher detection rates than maternal serum screening tests with minimal false positives and false negatives. Normal results could be very reassuring with proper genetic counseling
- More comprehensive than maternal serum screening tests—can detect sex chromosome abnormalities in addition to trisomy 13, trisomy 18 and trisomy 21. SNP-based NIPT can also detect triploidy and common micro-deletion syndromes

Table 6.1: Comparison of clinical validation data of different NIPT companies*

Chromosomal abnormality	Sequenom materni T21[TM 31-33]		Verinata verifi[TM 26,34]		Arlosa harmony[TM 35-37]		Natera panorama[TM 38-41]	
	Sensitivity	False positive rate	Sensitivity	False positive rate	Sensitivity	False positive rate	Sensitivity	False positive rate
Trisomy 21 (Down syndrome)	99.1%	0.2%	99.9%	0.2%	100%	0.1%	100%	0%
Trisomy 18 (Edward)	100%	0.4%	97.3%	0.4%	98%	0.1%	100%	<0.1%
Trisomy 13 (Patau syndrome)	91.7%	0.1%	87.5%	0.1%	80%	0.05%	100%	0%
Triploidy	Unable to detect		Unable to detect		Unable to detect		100%	0%

* False positive rates are much lower than would be expected since these populations have higher disease prevalence than groups reflecting a normal distribution of maternal age

Table 6.2: Summary on counting NIPT *vs* SNP-based NIPT

	Quantitative NIPT (Sequenom™, Verinata™, Ariosa™ Beijing Genomic Institute)	Quantitative NIPT (Natera™)
Fetal cell free DNA distinguished from maternal cell free DNA	No—Counts the total DNA from a chromosome (maternal and fetal) and compares to a reference chromosome.[14,15]	Yes—Provides increased sensitivity and can also detect maternal mosaicism reducing false positives.[14,15,21,41]
Sensitivity for common chromosomal abnormalities	T21 ->99%	T21 ->99%
Sensitivity for common chromosomal abnormalities	T18 –97.4% – >99% T13 –80% – 91.7%	T18 ->99% T13 ->99%
Sensitivity for Down syndrome at low fetal fraction (4–8%)	75%[5]	>99%[42]
Fetal fraction reported	Some do not evaluate and most do not report	Yes
Multiple gestation	Yes	Under validation
Ability to detect triploidy	No	Yes
Ability to detect vanishing twin	No >15% false positive in discordant twins[24]	Yes
Microdeletion syndromes	Not tested	DiGeorge, Angelman, Prader-Willi, Cri-du-Chat, 1p36 deletion

- Convenient test requiring a simple blood draw from pregnant woman.

Limitations

- High risk results need to be confirmed with invasive prenatal diagnosis
- Cannot detect open neural tube defects and provide other pregnancy related information like risk for pre-eclampsia as possible with maternal serum screening tests
- Cannot detect atypical chromosomal abnormalities (those other than aneuploidy, triploidy and copy number variants) like translocation, inversion, etc.
- Has a no call or no result rate of approximately 4%
- Small proportion of women may receive indeterminate results with grey areas like "aneuploidy suspected" with the quantitative NIPT method.

Non-invasive Prenatal Testing in Clinical Practice

American College of Obstetricians and Gynecologists (ACOG) and other professional bodies including Society of Maternal-Fetal Medicine (SMFM), International Society of Prenatal Diagnosis (ISPD) and American College of Medical Genetics (ACMG) have provided recommendations for use of NIPT in clinical practice.

According to ACOG committee opinion, NIPT should be offered to the following women:[43]

- Age >35 years at delivery
- USG findings indicating risk for aneuploidy
- History of prior pregnancy with trisomy
- Parent carries a balanced Robertsonian translocation with increased risk of trisomy 13 or trisomy 21
- Screen positive from maternal serum screening tests.

Given that the true positive rate for maternal serum screening (MSS) is only 5%, offering NIPT would help identify truly high risk women because of its superior detection rate.[4,5] This would help reduce unnecessary invasive diagnostic tests and their associated risk for fetal

loss. Additionally, NIPT can provide more information for women who do not want to undertake risks of invasive prenatal testing. This has been substantiated by a retrospective analysis of screen positive women offered NIPT in addition to invasive prenatal diagnosis. The invasive testing rate was reduced from 47.2% to 39.2%. More importantly, MSS screen positive women declining amniocentesis was reduced from 52.8% to 21.1%, when these women were also given the option to have NIPT.[44]

Various models to incorporate NIPT into routine prenatal screening also have been proposed to help reduce invasive prenatal diagnosis rates. One study evaluated a contingent screening model incorporating first trimester screening with NIPT. It demonstrated that invasive diagnosis rates can be significantly reduced to <0.5%.[45]

NIPT is a superior screening test but does not replace the diagnostic accuracy of invasive prenatal diagnosis. NIPT should be offered as an option and not as part of a regular prenatal testing protocol. Positive cases for NIPT should be offered invasive prenatal diagnosis for confirmation of test results. NIPT should not be offered in low risk pregnancies and multiple gestation pregnancies as it has not been adequately evaluated in these patient groups. NIPT cannot detect open neural tube defects and structural fetal anomalies. Hence, maternal serum alpha-fetoprotein (AFP) and ultrasound evaluation should continue to be offered. If a structural fetal anomaly is identified in ultrasound, invasive prenatal diagnosis should be offered because NIPT can only detect trisomy 13, trisomy 18, trisomy 21, monosomy X and triploidy.[43]

Genetic Counseling

Genetic counseling for women considering NIPT is challenging and very critical and should be offered both pre- and post-test.[43] Previous to the availability of NIPT, high risk women had to make a decision on whether to undergo invasive testing or not. With availability of NIPT and prenatal chromosomal microarray, women will now have to choose between NIPT, invasive testing with karyotyping, invasive testing with microarray and no testing. A detailed family and personal history should be obtained before testing to determine if the patient would more benefit from invasive testing or other forms of screening. These decisions can be significantly informed through discussion with a counselor. Women who screen positive from NIPT should be offered genetic counseling and options for invasive prenatal diagnosis for confirmation of test results as false positive results can occur. A low risk result does not ensure an unaffected pregnancy as false negative results can also occur. Also, the range of chromosomal abnormalities that can be detected by NIPT are fewer than by karyotyping (although changing rapidly) and significantly lower than by microarray.[43] These limitations need to be considered and explained in detail along with the limitations of other testing options of invasive diagnosis and chromosomal microarray. The scope, benefits and limitations of all testing options need to be carefully explained to enable women to make an informed decision.

Future Directions

Research efforts have been made in extending the scope of NIPT from whole chromosome abnormalities to sub-chromosomal abnormalities. Proof of principle that fetal sub-chromosomal abnormalities can be detected with NIPT was established in two studies in which the fetal karyotype from earlier invasive testing was known. In the first study, a paternally inherited microdeletion on chromosome 12 and in the second study, NIPT detected 22q11.2 microdeletion in two pregnancies with confirmed DiGeorge syndrome fetuses.[46, 47]

The Panorama™ test from natera has recently launched the microdeletion panel which screens for the most common and severe microdeletion syndromes in addition to the basic screening panel for trisomy 13, trisomy 18, trisomy 21, triploidy and sex chromosome abnormalities. The microdeletion panel screens for DiGeorge syndrome, Angelman syndrome, Prader-Willi syndrome, Cri-du-Chat syndrome and 1p36 deletion. Validation studies on 469 samples, including 110 confirmed positives showed sensitivity of 93.8–100% for these microdeletion syndromes with specificity of >99%.[48]

Future prospects for NIPT would include single gene disorders and sequencing of entire fetal genome. Although there have been studies to show that this is technically feasible, the clinical utility of this test is currently not known.[15] NIPT is revolutionizing prenatal screening and diagnosis and growing rapidly. These are truly exciting times in clinical laboratory medicine.

REFERENCES

1. Ferguson-Smith MA, Yates JRW. Maternal age specific rates for chromosome aberrations and factors influencing them: report of a collaborative European study on 52965 amniocenteses. Prenat Diagn 1984; 4:5–44.

2. Tabor A, Alfirevic Z. Update on procedure-related risks for prenatal diagnostic techniques. Fetal Diagn Ther 2010; 27:1–7.

3. Benn PA, Egan JF, Fang M, Smith-Bindman R. Changes in the utilization of prenatal diagnosis. Obstet Gynecol 2004; 103:1255–1260.

4. Cuckle H, Benn P. Multianalyte maternal serum screening for chromosomal defects. Genetic Disorders and the Fetus: Diagnosis, Prevention and Treatment (6th edn), In: Milunsky A, Milunsky JM (eds). Wiley-Blackwell, Chichester , UK, 2010; pp. 771–818.

5. Syngelaki A, Cheleman T, Dagklis T, Allan L, Nicholaides KH. Challenges in the diagnosis of fetal non-chromosomal abnormalities at 11–13 weeks. Prenat Diagn 2011; 31: 90–102.

6. Elissa Gussin, HA et Elias S. Culture of fetal cells from maternal blood for prenatal diagnosis. Human Reproduction Update 2002; 8:523–527.

7. Bianchi DW, Simpson JL, Jackson LG, et al. Fetal gender and aneuploidy detection using fetal cells in maternal blood: analysis of NIFTY 1 data. National Institute of Child Health and Development Fetal Cell Isolation Study. Prenat Diagn 2002; 22(7); 609–615.

8. Guetta E, Simchen MJ, Mammon-Daviko K, et al. Analysis of fetal blood cells in the maternal circulation: challenges, ongoing efforts and potential solutions. Stem Cells Dev 2004; 13(1):93–99.

9. Lo YM, Corbetta N, Chamberlain PF, et al. Presence of fetal DNA in maternal plasma and serum. Lancet 1997; 350:485–487.

10. Illanes S, Denbrow M, Kailasam C, Finning K, et al. Early detection of cell-free fetal DNA in maternal plasma. Early Hum Dev 2007; 83:563–566.

11. Lo YM, Tein MS, Lau TK, Haines CJ, et al. Quantitative analysis of fetal DNA in maternal plasma and serum: implications for non-invasive prenatal diagnosis. Am J Hum Genet 1998; 62:768–775.

12. Lun FM, Chiu RW, Allen Chan KC, Yeung Leung T, et al. Microfluidics digital PCR reveals a higher than expected fraction of fetal DNA in maternal plasma. Clin Chem 2008; 54:1664–1674.

13. Smid M, Galbiati S, Vassallo A, Gambini D, et al. No evidence of fetal DNA persistence in maternal plasma after pregnancy. Hum Genet 2003; 112:617–618.

14. Benn H, Cuckle H, Pergament E. Non-invasive prenatal testing for aneuploidy: current status and future prospects. Ultrasound Obstet Gynecol 2013; 42:15–33.

15. Bianchi DW, Wilkins-Haug L. Integration of noninvasive DNA testing for aneuploidy into prenatal care: What has happened since the rubber met the road? Clinical Chem; 60:78–87.

16. Palomiki GE, Kloza EM, Lambert-Messerlian GM, Haddow JE, et al. DNA sequencing of maternal plasma to detect Down syndrome: An international clinical validation study. Genet Med 2011; 11:913–920.

17. Canick JA, Palomiki GE, Kloza EM, Lambert-Messerlian GM et al. The impact of maternal plasma DNA fetal fraction on next generation sequencing tests for common fetal aneuploidies. Prenat Diagn 2013; 33:667–674.

18. Alamillo CM, et al. Nearly a third of abnormalities found after first-trimester screening are different than expected: 10-year experience from a single center. Prenatal Diagnosis 2013; 33:251–56.

19. Wang E, Batey A, Struble C, Musci T, et al. Gestational age and maternal weight effects on fetal cell-free DNA in maternal plasma. Prenat Diagn 2013; 33:662–666.

20. Bianchi DW, Platt LD, Goldberg JD, Abuhamad AZ, et al. Genome-wide fetal aneuploidy detection by maternal plasma DNA sequencing. Obstet Gynecol 2012; 119:890–901.

21. Wang Y, Chen Y, Tian Feng, Zhang J, et al. Maternal mosaicism is a significant contributor to discordant sex chromosomal aneuploidies associated with noninvasive prenatal testing. Clin Chem 2014; 60:251–259.

22. Sehnert AJ, Rhees B, Comstock D, De Feo E, et al. Optimal detection of fetal chromosomal abnormalities by massively parallel DNA sequencing of cell-free fetal DNA from maternal blood. Clin Chem 2011; 57:1042–1049.

23. Lau TK, Jiang F, Chan MK, Zhang H, et al. Non-invasive prenatal screening of fetal Down syndrome by maternal plasma DNA sequencing in twin pregnancies. J Matern Fetal Neonatal Med 2013; 26:434–437.

24. Futch T, Spinosa J, Bhatt S, de Feo S, et al. Initial clinical laboratory experience in noninvasive prenatal testing for fetal aneuploidy from maternal plasma DNA samples. Prenat Diagn 2013; 33:569–574.

25. Chen CP, Chien SC et Lin HH. Prenatal sonographic features of Triploidy. J Med Ultrasound 2007; 15:175–182.

26. Bianchi DW, Platt LD, Goldberg JD, Abuhamad AZ, et al. On behalf of the maternal blood is source to accurately diagnose fetal aneuploidy (MELISSA) study group. Genome-wide fetal aneuploidy detection by maternal plasma DNA sequencing. Obstet Gynecol 2012; 119:890–901.

27. Seckl MJ, et al. Choriocarcinoma and partial hyatidi form moles. Lancet 2000; 356:36–39.

28. Chromosome Abnormalities and Genetic Counseling, Gardner and Sutherland, 2004.

29. International Human Genome Sequencing Consortium. Initial sequencing and analysis of the human genome. Nature 2001; 409:860–921.

30. Chen YC, Liu T, Yu CH, Chiang TY, et al. Effects of GC bias in Next Generation Sequencing Data on *de novo* genome assembly. PLoS one 2013:8; e62856.

31. Palomiki GE, Deciu C, Kloza CM, Lambert-Messerlian GM et al. DNA sequencing of maternal plasma reliably identified trisomy 18 and trisomy 13 as well as Down syndrome: an International Collaborative Study. Genet Med 2012; 3:296–305.

32. Sequenom Internal Data (www.sequenom.com)

33. Mazloom AR1, D•akula •, Oeth P, Wang H, et al. Noninvasive prenatal detection of sex chromosomal aneuploidies by sequencing circulating cell-free DNA from maternal plasma. Prenat Diagn 2013; 33:591–597.

34. Verinata Internal Data (www.verinata.com)

35. Ashoor G, Syngelaki A, Wagner M, Birdir C, et al. Chromosome-selective sequencing of maternal plasma cell-free DNA for first-trimester detection of trisomy 21 and trisomy 18. Am J Obstet Gynecol 2012; 206. 322 e. 1–5

36. Ashoor G, Syngelaki A, Wang E, Struble C, et al. Trisomy 13 detection in the first trimester of pregnancy using a chromosome-selective cell-free DNA analysis method. Ultrasound Obstet Gynecol 2013; 41:21–25.

37. Ariosa Internal Data (www.ariosa.com)

38. Levy B, et al. Highly multiplexed targeted single-nucleotide polymorphism (SNP) amplification and sequencing as a method for identifying fetal chromosomal disorders from maternal cell-free DNA. Presented at ESHRE 2013.

39. Nicolaides KH, Syngelaki A, Gil M, Atanasova V, et al. Validation of targeted sequencing of single-nucleotide polymorphisms for non-invasive prenatal detection of aneuploidy of chromosomes 13, 18, 21, X and Y. Prenat Diagn 2013; 33:575–579.

40. Samango-Sprouse C, Banjevic M, Ryan A, Sigurjonsson S, et al. SNP-based non-invasive prenatal testing detects sex chromosome aneuploidies with high accuracy. Prenat Diagn. 2013; 33(7):643–649.

41. Natera Internal data (www.panoramatest.com)

42. Rabinowitz M, Valenti E, Pettersen B, Sigurjonsson S. Noninvasive aneuploidy detection by multiplexed amplification and sequencing of polymorphic Loci. Obstet Gynecol. 2014; 123 (Suppl 1):167S.

43. ACOG committee opinion no. 545, 2012.

44. Chetty S, Garabedian MJ, Norton ME. Uptake of noninvasive prenatal testing (NIPT) in women following positive aneuploidy screening. Prenat Diag 2013; 33:542–546.

45. Nicolaides KH, Wright D, Poon LC, Syngelaki A, et al. First-trimester screening for trisomy 21 by biomarkers and maternal blood cell-free DNA testing. Ultrasound Obstet Gynecol 2013; 42:41–50.

46. Peters D, Chu T, Yatsenko SA, Hendrix N, et al. Non-invasive prenatal diagnosis of a fetal microdeletion syndrome. N Engl J Med 2011; 365:2175–2184.

47. Jensen TJ, Dzakula Z, Deciu C, Van den Boom D, et al. Detection of microdeletion 22q11.2 in a fetus by next-generation sequencing of maternal plasma. Clin Chem 2012; 58:1148–1151.

48. Natera internal data for microdeletions.

7

Prenatal Diagnostic Procedures for High Risk Pregnancies

GENERAL GUIDELINES FOR INVASIVE PROCEDURES

Transabdominal-chorion villous sampling (TA-CVS), amniocentesis and FBS (Fig. 7.1) share a number of methodological criteria: ultrasound examination, skin disinfection,

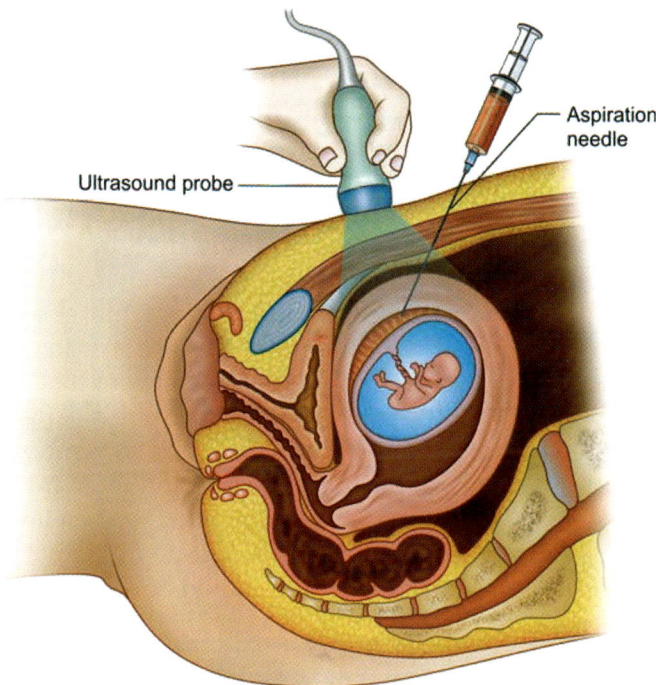

Fig. 7.1: Transabdominal CVS

percutaneous needling by a 20 to 22 gauge, 9 to 12 cm in length, spinal needle, continuous ultrasound visualization of the sampling device. The preliminary ultrasound study provides a clear anatomical view of the uterus and its relationship with the bladder and the bowels, myomas and placental abnormalities or location; multiple pregnancy and fetal abnormalities are also investigated. In each procedure transabdominal needling can be most safely performed by a two-person team (ultrasonographer and sampler): this way the needle path and progression through tissue can be identified and followed.

Sampling needles of more than 12 cm in length are too flexible to be correctly guided and thus should be avoided. The target should be firstly identified and its distance from the surface measured; bladder filling or voiding can make the target more easily reachable.

The use of a two-person team and the free-hand needle insertion allow the use of non-sterile ultrasound probe, that stays at few centimeter from the disinfected field, and this makes the whole procedure safer, quicker, cheaper and more appropriate for in-office patients than a double coaxial needle system inserted through a needle guide attached to the ultra-sound probe. In principle, convex ultrasound probe should be preferred to reliably guide the sampling device: its small contact surface allows an optimal fit to the maternal abdomen and facilitates an easy handling of the transducer in all the scanning sections, with a comprehensive view of the uterus still frequently situated near or below the pubis. After the procedure, maternal hemostasis by light pressure should be provided and fetal well-being observed and shown to the mother.

When Rh immune prophylaxis is indicated, 100 to 200 µg of anti-D immune gammaglobulin should be administrated by intra-muscular injection in the very first hours to avoid any maternal sensitization.

No more than two insertions should be attempted to obtain the fetal tissue: an increased fetal loss rate has been

documented for CVS and amniocentesis (17) when more than two attempts have been performed. In case of sampling failure any further attempt should be postponed, the patient informed and the methodology eventually reconsidered.

Chorionic Villous Sampling

Chorionic villous sampling (CVS) for confirmation in screen positive cases has emerged as the logical choice. The chorionic tissue is suitable for cytogenetic tests, DNA testing as well as metabolic studies. Though amniocentesis is the most commonly performed prenatal diagnostic procedure CVS has the advantage of early and rapid diagnosis in first trimester thus reducing anxiety, keeping the pregnancy private and also to offer the safer first trimester termination option. To achieve the high safety and success of CVS it is essential to have properly trained operator with experience of at least 300–400 tests and should be well versed in embryo-logy and first trimester USG. Good resolution ultrasound machine also makes the procedure quick and safe.

Clinical Procedure: TA-CVS

Two main approaches are available for obtaining chorionic tissue: The placenta is reached either through the maternal abdominal wall by needling (TA-CVS) or through the vagina and the cervical canal by the insertion of a CVS cannula (developed by the author). When the procedures have been performed by operators well experienced in both approaches, no fetal loss rate differences were observed between TA- and TC-CVS, and both approaches were equally successful. It is necessary for the clinician to be well versed in both these techniques and use them depending upon the need. Certain advantages have been claimed by those who prefer the TA approach, they are, (a) lower rate of complications (bleeding, uterine infection), (b) less maternal contamination, (c) sampl-ing feasibility and safety beyond the first trimester (late booking, diagnosis after the first as well as the second trimester screening) in alternative to early or mid-trimester amniocentesis.

The free-hand transabdominal needle insertion is a four steps procedure: (1) the insertion point on the maternal abdominal surface and the sampling pathway are chosen in relation to the distance of the target (the placenta) as well as to the length of the needle available; then (2) the needle is inserted through abdominal wall and stopped precisely next to the external uterine surface, and (3) with earlier confirmation of the correct direction the needle is pushed further into the principle axis of the placenta; (4) the stylet is removed and a 10 ml syringe with 1 ml of saline solution/CVS medium is attached, and repeated suctions by careful backward and forward movements of the needle are performed for about 20 seconds. The sampling system is withdrawn by keeping a fully depressed syringe, and chorionic tissue put in a petri dish for immediate confirmation, washing, and selection under dissecting microscope. The tissue specimen can be preserved in culture medium at room temperature (about 20°C) to be processed by the cytogenetic laboratory in 24–30 hours: in case DNA analysis is requested a tissue aliquot can be preserved in Eppendorf tubes with saline solution at 4°C for several days.

Clinical Procedure: TC-CVS (Figs 7.2A and B)

Detailed pelvic ultrasound examination is essential to determine:

- Viability, accurate dating
- Number of gestation sacs, chorionicity (in case of multiple pregnancies)
- Implantation of placenta, presence of other complicating factors like fibroid, low implantation, sub-chorionic bleeding
- Uterine position, utero-cervical angulation
- Fullness of bladder

TC route is particularly useful when placental implantation is on posterior uterine wall, lower uterine areas or presence of fibroids on the anterior surface of the uterus. Steps involved are:

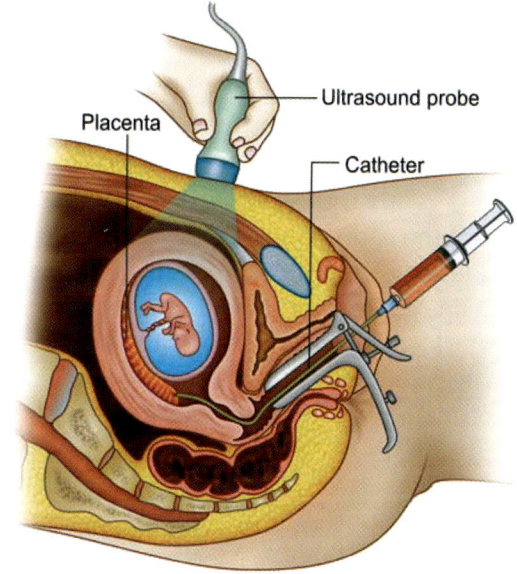

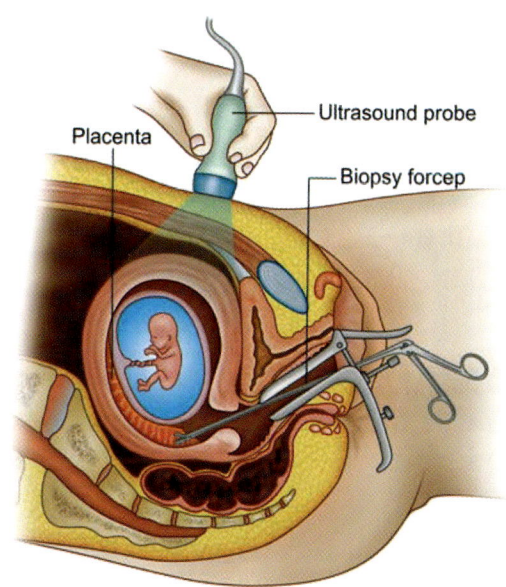

Figs 7.2A and B: Transcervical CVS

1. Proper vaginal toilet to disinfect vagina, cervical canal.
2. Speculum examination, stabilizing the cervix by long Alles forceps.
3. CVS cannula is moulded according to cervicouterine angulation and placental site, cannula is flushed with CVS medium and 1cc medium is aspirated.
4. Cannula is gently introduced in the cervical canal, further passage of the cannula is guided by the USG scanning.
5. Once the tip of the cannula reaches lower border of the placenta it is gently introduced in the substance along the length of placenta halfway along the length.
6. Once in position suction is created (3–4 cc by withdrawing the piston) cannula is gently moved to-fro for 15–20 seconds.
7. It is important to release the suction before withdrawing the cannula.
8. Contents of the syring and cannula are discharged in a petri dish containing CVS medium, chorionic villi are picked up with forceps, cleaned of mucus, deciduas rinsed in tissue culture medium and transferred into sterile tube containing transport medium.

Post procedure USG scan is done to look for any sub-chorionic bleed and to reassure the patient.

Major clinical experiences in first trimester CVS is shown in Table 7.1.

Table 7.1: Clinical experience of CVS					
Author	Technique	No. of cases	Gest weeks	Success loss %	Preg. (%)
1. Brambati, et al.	TC	1305	8–12	99.2	3.9
2. Brun et al	TA	10741	8–38	99.9	1.7
3. Jackson and Wapner	TA	11600	9–12	99.7	1.9
4. Williams, et al.	TC	2949	9–12	99.7	1.9
5. Brambati, et al.	TC/TA	10000	8–32	99.7	2.6
6. Gogate, et al.	TC/TA	10050	9–28	98.8	2.1
(References 1–6)					

Doubt about the safety of CVS was raised after the cluster of severe limb reduction defect anomalies (LRD) reported by Firth, et al. in 1991. WHO statement based on the complete follow-up of 76476 cases by the WHO-CVS registry did not show any increased risk of LRD following CVS performed by well trained operators and performed only after eighth completed week of pregnancy. The risk of bleeding, feto-maternal hemorrhage, infection and pregnancy loss is more than amniocentesis but due to it's advantage of early diagnosis the test is becoming more popular. *Brambati B, Tului L*[7].

Fetal Tissue Sampling in Second Trimester

In spite of the screening tests, ultrasonography and other non-invasive tests certain high risk patients have to undergo fetal tissue sampling for confirmation and prognostication of the fetal disorders like chromosomal anomalies, single gene disorders, IEM, fetal infections, etc. To carry out these interventional tests reliably and safely we have to have genetic centers of excellence with well trained clinical and laboratory facilities, multi-specialty fetal medicine program.

Amniocentesis (Fig. 7.3) the most prevalent test and is considered as a gold standard. Genetic amniocentesis is conventionally done between 15–24 weeks of pregnancy. The amniotic fluid is tested for chromosomal evaluation by study of cultured/uncultured amniocytes (conventional karyo-typing, FISH/QF-PCR), study of single gene disorders by DNA diagnostic tests and evaluating errors of metabolism by study of supernatant fluid or amniocytes. Fetal involvement in maternal infections can be verified by DNA testing, immunological work-up.

Clinical Procedure

Amniocentesis is performed through the maternal abdominal wall by a 22 gauge (0.7 mm in diameter), 9–12 cm in length spinal needle. The needle-insertion site on the abdominal surface should respect the following criteria:

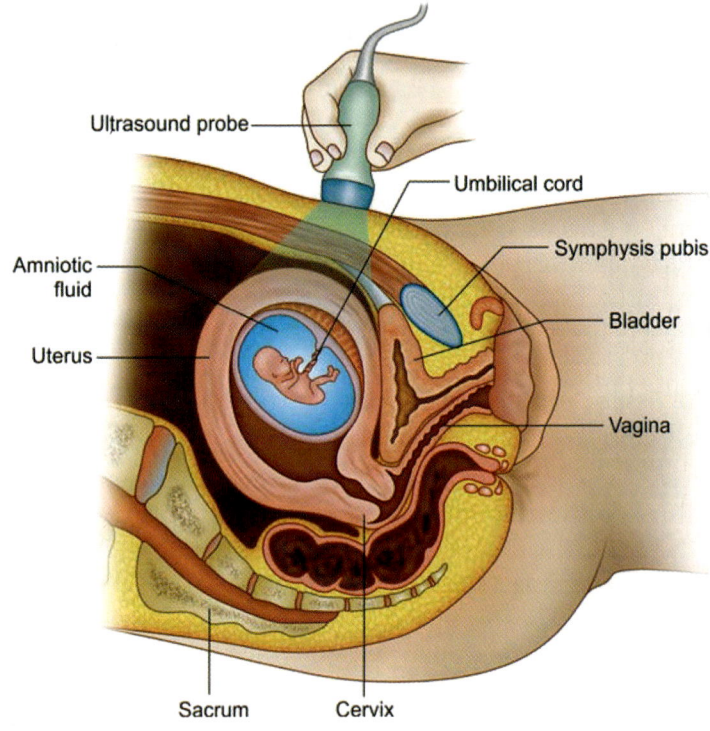

Ultrasound probe

Umbilical cord

Symphysis pubis

Amniotic fluid

Bladder

Uterus

Vagina

Sacrum   Cervix

Fig. 7.3: Amniocentesis

a. It should be as close as possible to the midline fundal area (to avoid puncturing large vessels), and on the fetal limb side where an optimal pocket of AF can visualized (to avoid damages to the major fetal structures).

b. It should be directed perpendicular to the inner uterine surface and inserted by a determined and well controlled movement (to avoid pushing back amniotic membrane).

c. Placenta should be avoided (because of the potential damage of fetal vessels on the chorionic plate).

d. The syringe should be removed before pulling out the needle (to avoid any maternal tissue contamination).

e. All the previous steps should be monitored under continuous ultrasound control. After assurance that the needle

is in its proper location, the stylet is moved, a 10 ml disposable syringe attached and 15–20 AF ml aspirated, AF is then transferred in a labeled sterile flask or test-tube and kept at room temperature (about 20°C) to be sent to the laboratory within 24–48 hours.

The clinical safety and success of amniocentesis depends upon:

1. Experience of the operator performing the procedure.
2. Characteristics of the amniotic fluid(blood staining/ discolored fluid).
3. Direct ultrasonography supervision.
4. Indication for the procedure *Elias S, Simpson JL*[8]. The clinical efficacy of this test performed in over 500 cases last year by our center was success rate of 99.6% and pregnancy loss rate of 0.4%. The efficacy of the laboratory testing was over 99% with failure to grow AF culture in less than 1% (*Gogate et al.*[9])

Fetal blood sampling for prenatal diagnosis is most likely to be done for rapid fetal karyotyping, evaluation of fetal hematological disorders and fetal infections, DNA diagnosis. Therapeutic procedures like intrauterine transfusions, drug therapy and stem cell transfer are also performed through cordocentesis. Direct ultrasound guided cord blood sampling is the most commonly preferred technique. Safety of the procedure is acceptable with uncorrected pregnancy loss rate of 1.6% in a large collaborative data from 14 North American Centers (*Weiner CP, Okamura K*[10]). In a similar multi-centric study by the author, 1216 cordocentesis were evaluated from seven centers in India with success rate of 93.2% and uncorrected pregnancy loss of 2.3% (*Gogate SG, et al.*[11]).

The extension of transabdominal CVS in second and third trimester is a very convenient, safe alternative to amniocentesis and fetal blood sampling particularly as an alternative to early amniocentesis, when patient presents late or an anomaly diagnosed beyond 26 weeks or when amniocentesis or cordocentesis is confounded by unfavorable conditions like severe oligohydramnios *(Holzgreve W, Miny P, SchlooR, et al.[12]), (Podobnik M, Siglar S, Singer Z, et al.[13])*. We have also been using late CVS in second and third trimesters as an alternative to fetal blood sampling and amniocentesis with gratifying results.

REFERENCES

1. Gogate SG. Preventive Genetics-Holistic healthcare; Preventive Genetics, Gogate SG, (Ed.). Jaypee Brothers, New Delhi, 2006; pp. 3–24.

2. Down's syndrome screening programme for England—A Handbook for Staff. 2004. Ward PA, Gray JA. M. ISBN 09543684-1-X. National Screening Committee Programmes Directorate.

3. Nicolaides HN, Sebire J, Snijders JM. The 11–14 week scan. The diagnosis of fetal abnormalities. 1999; pp. 1467–2162

4. Spencer K, Souter V, Tul N, Snijder R, Nicolaides KH. A screening program for Trisomy 21 at 10–14 weeks using foetal nuchal translucency, maternal serum free beta-HCG and PAPP-A. Ultrasound ObstetGynecol 1999;13: 231–7.

5. Milunsky A, Canick JA. Maternal serum screening for neural tube and other defects. Genetic disorders and the foetus, Milunsky A (Ed). John Hopkins University press; 2004: pp. 719–94.

6. Canick JA, Knight GJ, Palomaki GE. Low second trimester maternal serum unconjugated oestriol in pregnancies with Down's syndrome. Br J Obstet Gynaecol 1988; 95: 330.

7. Bogart HM, Pandian MR, Jones OW. Abnormal maternal serum chorionic gonadotropin levels in pregnancies with foetal chromosomal anomalies. Prenat Diagn 1987; 7:623.

8. Morris JK, Wald NJ. Quantifying the decline in the birth prevalence of neural tube defects in England and Wales. J Med Screening 1999; pp. 6–182.

9. Knight GJ, Maternal serum alpha protein screening techniques in diagnostic human biochemical genetics, (Ed.). Hommes FA, Weily-Liss, 1991:491.

10. Cuckle HS, Arbuzova S. Multi marker maternal serum screening for chromosomal abnormalities. Genetic disorders and the foetus, (Ed.). Milunsky A, John Hopkins University press 2004; 795–835.

11. Wald NJ, Bestwik JP, Morris JK. Cross trimester ratios in prenatal screening for Down's syndrome. Prenat Diagn 2006; 26:514–523.

12. Bianch DW. From Michael to Microarrays: 30 years of studying fetal cells in maternal blood. Prenat Diagn 2010; 30:622–623.

13. Mouawia H, Saker S, Jais JP, et al. Circulating trophobalstic cells provide genetic diagnosis in 63 fetuses at risk for cystic fibrosis, spino-muscular atrophy. Reproductive biomed online 2012; 25(5):508–520.

8

Genetic Confirmatory Test After Maternal Serum Screening Test

Hema Purandarey

INTRODUCTION

In the early years, major indication for invasive prenatal diagnosis for Down's syndrome was advanced maternal age (age >35 yrs). Since mid 90s,[1] maternal serum screening (MSS) was introduced for prenatal screening in general pregnant population.

Over the years, more parameters were added in the prenatal screening in second as well as first trimester and are today accepted as a screening test in women even younger than 35 yrs.[1,2] These parameters in addition of USG markers has increased the risk prediction for Tri 21, 18 and NTD to a greater extent and its application in elderly mothers have also given an opportunity to select invasive testing with more informative risk, reducing the very small loss of normal fetuses with prenatal diagnostic procedures.

Present chapter discusses the impact of MSS and laboratory confirmatory tests by amniocentesis or chorionic villus sampling (CVS) in high risk cases.

CURRENT SCENARIO OF MATERNAL SERUM SCREENING AND REVIEW

Maternal serum screening today is a worldwide accepted method for identifying women at an increased risk for fetal chromosomal aneuploidies mainly trisomy 21, 18 and neural tube defects (NTD). The calculated risks identified are above general population risk and maternal age associated risks.

Markers used are pregnancy associated plasma protein A (PAPP-A), free β-human chorionic gonadotropic (fβ-hCG), alpha fetoprotein (AFP), unconjugated estriol (uE3), inhibin (Inh-A) which are combined into dual (PAPP-A, fβ-hCG), triple (AFP, fβ-hCG, uE3), and Quadruple (AFP, fβ-hCG, uE3, Inh-A) screen as their efficacy lies in combined results. The mathematical calculation involves the levels of these substances after considering other factors namely maternal age, date of birth, weight, diabetes status, previous pregnancies with trisomy 21, number of fetuses, smoking status, ethnic origin, type of pregnancy (natural/induced/IVF). Inclusion of nuchal translucency (NT) finding in double marker test, increases the accuracy of the test (Table 8.1). The main objective of the tests is to offer pre-test genetic counseling; for effective diagnostic and therapeutic options, having a child who would have special needs and explain the fundamental difference between the screening and the diagnostic tests.

Review of current scenario from few countries and our experience regarding the uptake of its diagnostic ability and impact on Down syndrome outcome are discussed below. Overall increase in women of elderly age group is increased due to more couples starting or spacing birth of a child and possibility of parenthood by way of reproductive technology for infertile couples even at a later age.

Introduction of MSS and its impact on the birth prevalence of Down syndrome, follow-up of invasive procedures like CVS and amniocentesis is widely studied. A population based study in Australia has shown the uptake of prenatal testing went up from 7% (which was mainly for elderly woman undergoing CVS/amniocentesis) to 84%. Maternal serum testing followed by invasive procedures has shown that about half the terminations were in younger women.[3] Similar studies on 26,488 antenatal and postnatal diagnoses of Down syndrome made in cytogenetic laboratories in England and Wales in the year 2001, the UK national committee advised that all pregnant mothers should be

Table 8.1: Maternal serum screening (right test, right time)

Option	First trimester	First and second trimester combined		Second trimester	
	First screen	Sequential screen	Integrated screen	Serum integrated screen	AFP4
Benefits	The highest detection rate in the first trimester	Early answer and high detection rates	The highest detection rates	The highest detection rates without NT	The best second trimester screening
Down syndrome detection rate	83%	90.4%	92%	87%	81%
FPR	5%	3.7%	5%	5%	5%
OAPR	1 in 3	1 in 7/1.16	1 in 21	1 in 22	1 in 23
Trisomy 18 detection rate	80%	90%	90%	90%	80%
ONTD detection rate	—	80%	80%	80%	80%
Marker	NT, PAPP-A, hCG	NT, PAPP-A, hCG, AFP, hCG, Ue3, inhibin	NT, PAPP, AFP, hCG, Ue3, inhibin	PAPP, AFP, hCG, Ue3, inhibin	AFP, hCG, Ue3, inhibin
Timing	10–13 weeks	10–13 weeks 10–13 weeks	15–21 weeks 15–21 weeks	10–13 weeks 15–21 weeks	15–21 weeks

offered one of the screening tests for Down syndrome.[4] The paper published on the similar subject by the group states— "the number of diagnoses of Down syndrome has increased by 71% (from 1075 in 1989/90 to 1843 in 2007/08), whereas that of live births decreased by 1% (755 to 453), owing to antenatal screening and subsequent terminations.[5] In the absence of antenatal screening and subsequent terminations, the numbers of Down syndrome births would have increased by 48% due to parents choosing to start families later. A study on similar lines was conducted in Turkey to document the clinical and cytogenetic results of amniocentesis, wherein they concluded, abnormal MSS result is the most frequent indication and Down syndrome is the most commonly detected abnormality. Other major chromosomal abnormalities were accounting for 3.2%.[6]

At our center for genetic health care, between 2007–2014, a total of 4619 karyotypes were analyzed for various indications like maternal age (1928 cases) (41.7%), high risk in MSS (1503 cases) (32.5%), ultrasonography marker (635 cases) (13.7%) and others (705 cases) (15.2%). Of the total MSS test, 124 (8.25%) cases showed chromosomal abnormality. The abnormalities found were

Total maternal serum screening cases	1503
Total number of abnormalities	124
Autosomal aneuploidy +21 (55), +18 (10), +13 (01)	66
Sex chromosomal aneuploidy (XXX, XYY, XO)	04
Translocations – Robertsonian (5), reciprocal (11)	16
Inversions	38

Cynthia, et al. stated that a comprehensive counseling should be made available to all pregnant women.[5] Their study states though antenatal screening will increase the odds of identifying an abnormal fetus, by reducing the need of invasive diagnostic tests, the disadvantage of aneuploidy fetuses being not identified still prevails.

Screening tests are generally performed on healthy patients and are often offered to the entire relevant population. They are affordable, easy to use and interpretable. Their function is only to help define high risk group among the general population.

Diagnostic tests are generally complex and require sophisticated analysis and are performed on at risk population. The test results give definitive answer. Occasionally the results show unexpected findings or variants and needs experienced person for interpretation as the decisions taken then are irreversible. Tests are usually expensive and are performed only on high risk group.

Association between increased risk for chromosomal disorders with low maternal serum alphafetoprotein mainly Down syndrome was first studied by Merkatz et al. in the year 1984. Its association with Trisomy 18 as well is observed in subsequent years.

In the year 1988 Wald, et al. suggested that like alpha fetoprotein (AFP), beta human chorionic gonadotropin (β- hCG) and unconjugated estriol (uE3) which significantly increased the detection frequency of Down syndrome to approximately 60% of the total incidence.[8] Dimeric inhibin A is an additional marker that may raise the sensitivity by 3–7% for a given screen positive rate. This is a quadruple marker test. Currently therefore we have four screening methods. Double, triple and quadruple test. The appropriate use and timing can have different predictive values (Table 8.1).

DIAGNOSTIC TESTING

Couples who are at increased risk, need prenatal genetic counseling for confirmation by cytogenetic test which involves invasive procedures. Currently two commonly used methods are chorionic villus sampling and amnio centesis.

Prenatal genetic counseling after screen positive results is many a times determining factor in high risk cases these include testing versus non-testing, methodology safety

efficacy, cost benefits, impact on future pregnancies, options available and informed written consent.

In today's scenario, high risk pregnancy patients from remote areas also can avail the laboratory diagnostic facilities on transported specimens from distant obstetric units.

For an accurate laboratory diagnosis appropriate specimen is the key. The invasive procedures are thus briefly mentioned here.

Amniocentesis at 16–18 weeks for testing of Down syndrome has become a routine since its introduction in the late 1960s. It became possible in the year 1984 for first trimester (11–13 weeks) aneuploidy test with the introduction of chorionic villus sampling done either by transcervical or through transabdominal route.[9] Fetal blood sampling at 18–20 weeks was another method used for rapid karyotyping which slowly reduced due to introduction of fluorescent *in situ* hybridization (FISH) which was used for rapid diagnosis of common aneuploidies. All invasive techniques are undertaken after careful prenatal counseling and inform written consent and are voluntary.

CHORIONIC VILLUS SAMPLING

Chorionic villus sampling (CVS) was introduced into medical practice in 1984. The advantage of CVS is availability of early results. CVS is usually done at an earlier stage of gestation than amniocentesis and hence more embryos and fetuses with highly lethal condition can be found at this stage of pregnancy.

The chorion consists of an outer layer of trophoblast and an inner layer of synctiotrophoblast with a mesenchymal core containing blood vessels (Fig. 8.1). The chorionic villi surrounds the embryo. The ideal time for CV is 10–12 weeks. CVS is utilized to determine the chromosomal, enzymatic or molecular genetic status of the fetus. The types of culturing for cytogenetic studies in chorionic villus sampling are short term and long term cultures.

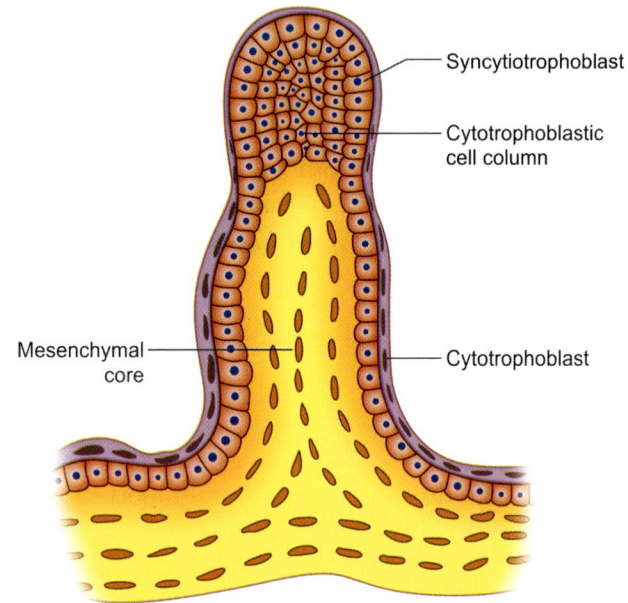

Fig. 8.1: Illustration of a typical chorionic villus (vertical section) showing the constituent layers

Short Term Culture

The principle behind the short term culture is to use the rapidly dividing cytotrophoblastic cells which form the outer covering of the villi. The results obtained are rapid (24–48 hrs) and there is less chance of maternal contamination however disadvantage is that chromosome morphology is poor and chances of tissue mosaicism is higher.

Long Term Culture

Long term culture has the advantage over short term in better morphology and very low chance of tissue mosaicism. However, chances of maternal cell contamination is more likely. Confined placental mosaicism appears in approximately 1% of the CVS specimens and in such cases, amniocentesis should be offered. However normal amniocentesis result does not eliminate the possibility of true mosaicism.

Identified confined placental mosaicism is associated with poor perinatal outcome, including increased risk of pregnancy loss, fetal growth restriction and stillbirth.

AMNIOCENTESIS

It is one of the more commonly performed procedures for prenatal diagnosis of fetal chromosomal aneuploidy. Types of fetal cells seen in the amniotic fluids are cells shed from fetus skin, amnion, the gastrointestinal and the genitourinary tract. Amniotic fluid obtained at 16–18 weeks of gestation has 30–80% of live cells. As the gestation increases, the total number of fetal cells increase but the viability of cells start decreasing.[10,11] There is less than 1% possibility of culture failure in amniotic fluid studies. There is also a rare possibility of maternal cell contamination which could lead to false diagnosis.

CORD BLOOD SAMPLING

It is one of the methods used to confirm the abnormal amniotic fluid karyotype. Cord blood sampling (CBS) can be done at advanced stages of gestation. Fetal blood cells can be cultured similar to lymphocyte culture and good karyotype results can be obtained in 24–72 hrs.

INTERPRETATION OF EXPECTED AND UNEXPECTED CYTOGENETIC RESULTS

Though majority of the invasive fetal tissue sampling world-wide is done for detection of Down syndrome many a times unexpected results are obtained which needs discussion here.

Interpretation of karyotype result is the most important aspect in prenatal diagnosis. The results obtained need to be correlated with the clinical indication in both expected and unexpected findings. An experienced cytogeneticist and clinical geneticist or fetal medicine expert should interpret them as the decision of continuation or termination of pregnancy is entirely based on the results obtained and the action taken is irreversible. Clinical significant abnormalities involving common aneuploidies (13, 18, 21 and sex chromosomes) Figures 8.2 to 8.4, contributes to approximately 80%. The constitution of the remaining 20% includes translocations (balanced and unbalanced), deletions, presence of a marker chromosome or more commonly found heteromorphic variation in general population. Identifying the chromosome involved by standard cytogenetic/molecular FISH studies can help in finding out the prognosis.

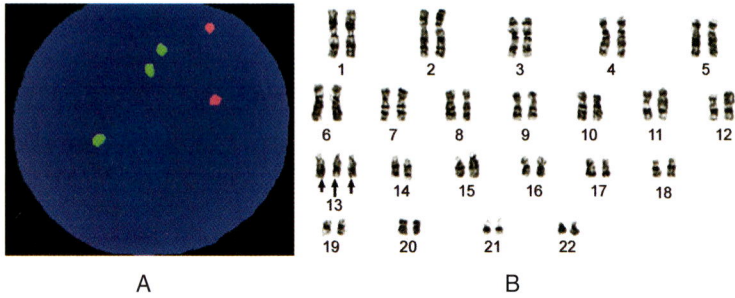

A   B

Fig. 8.2: (A) FISH image showing two orange signals for chromosome 21 and three green signals for chromosome 13 indicating trisomy 13, (B) Karyotype image of the same patient showing trisomy 13

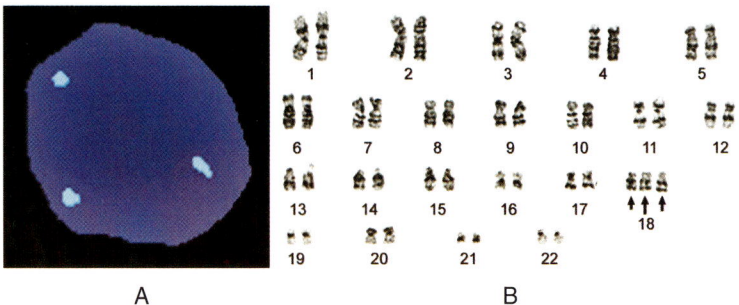

A   B

Fig. 8.3: (A) FISH image showing three aqua signals for chromosome 18 indicating trisomy 18, (B) Karyotype image of the same patient showing trisomy 18

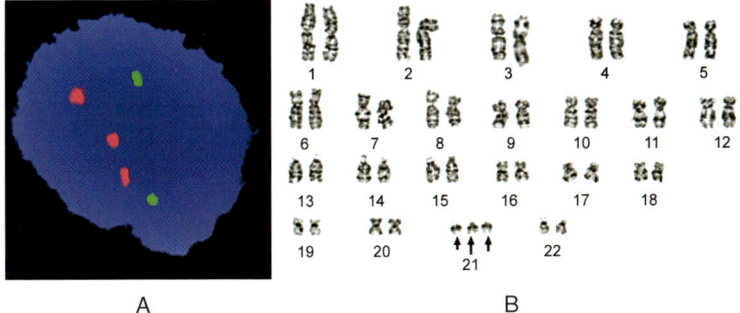

A   B

Fig. 8.4: (A) FISH image showing two green signals for chromosome 13 and three orange signals for chromosome 21 indicating trisomy 21, (B) Karyotype image of the same patient showing trisomy 21

Further confirmatory studies are required in prenatal cytogenetic findings such as balanced translocation, unbalanced translocation, marker chromosome or any heteromorphic variation (qh+, ps+, inversions). Parental cytogenetic studies in cases with heteromorphic variations is essential to find its origin, if *de novo* or familial. If familial, the couples are counseled accordingly and are give assurance about the normalcy of the fetus to be just like the carrier parent. If *de novo*, the couples are explained about the consequences which help them to make informed choice.

Reporting the origin of a marker chromosome is challenging. Marker being a chromosome of unknown origin requires different types of tests (FISH, C banding, etc.) to be performed to find its source. Markers may be present in a mosaic or a non-mosaic form. Whole chromosome paint gives an additional hand in identifying the origin of the marker chromosome (Fig. 8.5).

Turner syndrome (45, X) cytogenetic findings with normal ultrasound findings, requires a cautious approach to rule out mosaicism with normal cell line. A through discussion with the clinician of the ultrasound findings is recommended.

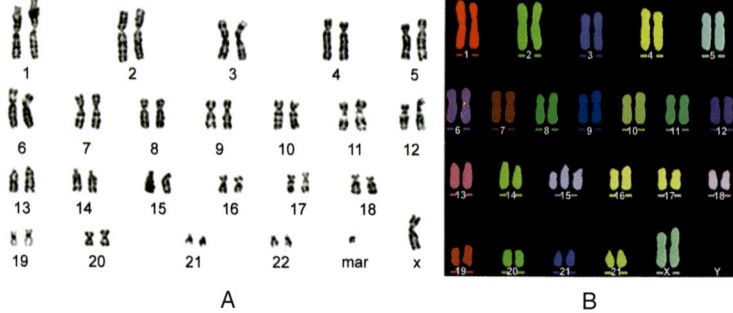

Fig. 8.5: (A) GTG banded male karyotype with presence of a marker chromosome, (B) Karyotype with M FISH stain identifies the marker as chromosome 15

Robertsonian translocation carriers involving the chromosomes 14 and/or 15 has a risk of having an offspring with uniparental disomy like Prader-Will syndrome and Angelman syndrome. In both inherited and *de novo* cases of Robertsonian translocations with chromosome 14 and/or 15, UPD needs to be ruled out.

MOLECULAR CYTOGENETICS (FLUORESCENT *IN SITU* HYBRIDIZATION)

With the inclusion of molecular techniques in the past 12 years, the diagnostic capabilities in prenatal and postnatal chromosome analysis have reached a height. The major inclusion being that of fluorescent *in situ* hybridization (FISH) technique which uses fluorescent labeled DNA probes which are hybridized to the metaphase spread of all the tissue types.

Interphase FISH in prenatal diagnosis helps in rapid assessment of the aneuploidy status on uncultured interphase cells from chorion villus sampling and amniotic fluid. The FISH studies are used to rule out the common aneuploidies (13, 18, 21 and sex chromosomes). Traces of blood in the sample may give no results in 2–5% of the samples.

Recent advances in QF-PCR and microarray with genomic clones are becoming popular and are promising of providing

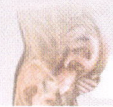

a replacement of FISH for microdeletion syndromes and subtelomeric analysis by providing details of complete genome which lacks in the G banding. The only disadvantage it has is the cost factor involved which give the FISH and conventional cytogenetic an upper hand.

Conclusion

The introduction of maternal serum screening has resulted in increased use of any prenatal testing for Down's syndrome. This has resulted in a significant fall in the birth prevalence of Down's syndrome. Further efforts to improve detection rates of aneuploidy with antenatal screening led to combining of existing first- and second-trimester screening. With combined testing, detection rates are improved. The goal of the screening is to identify the fetuses at high risk to have a congenital abnormality; after the screening they will be further investigated using invasive methods such as amniocentesis and chorionic villus biopsy. Confirmatory testing method allows option of selective termination of chromosomally abnormal fetus. In normal results, there is reassurence for rest of the pregnancy.

REFERENCES

1. Brock, Rodeck, Malcolm A. Ferguson-Smith. Prenatal Diagnosis and Screening. London: Longman Group UK Limited. 1992; 3–9, 13–21, 579–593.
2. Evans Johnson, Yaron Drugan. Prenatal Diagnosis. USA: McGraw-Hill Companies. 2006; pp. 415–421.
3. Cheffins T, Chan Annabelle. The impact of maternal serum screening on the birth prevalence of Down's syndrome and the use of amniocentesis and chorionic villus sampling in South Australia. British Journal of Obstetrics and Gynaecology. 2000; 107(–):1453–1459.
4. Joan Morris, Eva Alberman. Trends in Down's syndrome live births and antenatal diagnoses in England and Wales from 1989 to 2008: analysis of data from the National Down Syndrome Cytogenetic Register. *British Medical Journal.* 2009; 339 (–)
5. Cynthia Anderson, Charles Brown. Fetal Chromosomal Abnormalities: Antenatal Screening and Diagnosis. *AAFP* 2009; 79 (2), 117–123.

6. Zeynep Ocak, Mehmet Aygun. Clinical and cytogenetic results of a large series of amniocentesis cases from Turkey: Report of 6124 cases. *The journal of Obstetrics and Gynaecology Reserach.* 2014; 40 (–), 139–146.

7. Merkatz IR, Nitowsky HM, Macri JN, Johnson WE. An association between low maternal serum alpha-fetoprotein and fetal chromosomal abnormalities. Am J Obstet Gynecol 1984; 148:886–94.

8. Wald NJ, Cuckle HS, Densem JW, Nanchahal K, Royston P, Chard T, Haddow JE, Knight GJ, Palomaki GE, Canick JA. Maternal serum screening for Down's syndrome in early pregnancy. BMJ. 1988; 297(6653):883–887.

9. Smidt-Jenson S, Hahnemann N, Hariri J, Jensen PKA, Therkelsen AJ. Transabdominal chorionic villi sampling for first trimester fetal diagnosis: first 26 cases followed to term. Prenat Diagn. 1986; 6:125

10. Gosden CM, Brock DJH. Morphology of rapidly adhering amniotic fluid cells as an aid to the diagnosis of neural tube defects. Lancet I: 1977; 919–922.

11. Gosden CM, Brock DJH. Combined used of alphafetoprotein and amniotic fluid cell morphology in early prenatal diagnosis of fetal abnormalities. J Med Genet 1978; 15:262–270.

9

Overview of Expanded Maternal Serum Screening Program

How to Administer Maternal Serum Screening Program for NTD, Down's Syndrome?

- Awareness in medical fraternity and lay public about concept of screening program for Down's syndrome and neural tube defects. Establish close network with health care providers in public and private sectors so as to cover maximum target population.

- Conduct baseline studies in antenatal mothers at appropriate period of gestation to have local data on various markers and generate local median values and incorporate them in the screening software, standardize the screening tests.

- It is essential to have sample freezing facility to have normal as well as affected pregnancies as these samples can be used to quickly evaluate and setup newer screening tests.

- Facilities for patient education and pre- and post-test counseling, prepare patient education booklets, well designed clinical reference proforma, informed consent forms and laboratory report proforma.

- It is absolutely vital to prepare the sample referral proforma to cover all the essential parameters needed for giving optimum performance from the screening tests.

- It is essential to educate the clinicians as well as paramedical staff about proper completion of these proforma so as to ensure the desired specificity and sensitivity of the test. Incomplete proforma will make the risk assessment invalid and inaccurate thus defeating the basic purpose of the screening program.

- Ensure facilities for diagnostic tests, like cytogenetic testing for Down's syndrome and targeted USG scans and amniotic fluid tests for NTD, for high risk patients identified by the screening program.
- Establish facilities for counseling, management of the screened patients.
- Periodic work audits and assessment of the screening programs to improve the efficiency and data compilation, etc.

Likely Impact of a well Designed and Run Maternal Serum Screening Program for Down's and NTD

- More balanced attitude and increased awareness in the general population and health care workers.
- Better facilities for counseling, laboratory diagnostic tests and post-test management of affected pregnancies.
- Impact on Down's syndrome and NTD affected births.
- Better acceptance of pre-conception counseling, testing and management.

We sincerely hope that this handbook will be of some help in establishing this genetic screening program at all levels in all the parts of our entire country resulting better awareness, diagnosis and management of Down's syndrome and NTD affected pregnancies!

Annexures

- Information booklet for public
- Handbook for screening program staff
- Consent form
- Referral proforma
- Result proforma and clinical explanatory document
- Data compilation proforma

ANNEXURE 1: First Trimester Combined Screening Sample Referral and Patient Consent for First Trimester—Form*

Date:

Note: *Mandatorily to be filled

*Name: * DOB: Date:
*Address:
 Religion: Caste: *Age: * Weight:
*Clinician in charge:
E-mail ID:
Hospital address:
*Contact numbers:

Confounding factors:
*Pregestational diabetes: Yes/No   Smoking: Yes/No
Periconceptional folic acid: Yes/No
Previous chromosome abnormaly/NTD: Yes/No
If yes, specify:

Brief obstetric history:
* Gestational age by LMP: * LMP
* Sonography findings: * Scan date

Fetal status: Single / multiple =
(If multiple, specify for each fetus:

Gestational age (at date of scan):   *CRL:

*Nuchal translucency:

Date of blood collection (if different to date of ultrasound)

Any other information

Instructions for sample collection:

1. For first trimester screening (free β-hCG and PAPP-A) the sample should be collected between 11th to 14th (13th week and 6 days) week of pregnancy. The CRL must range from 45 to 84 mm

2. 5CC of blood must be collected in a plain bulb.

3. Serum can be stored for 6 days at 2–8°C.

4. Transportation should be done at 2–8°C. (In ice box)

Contact:

**Consent form for prenatal screening tests*

Govt. Reg. No. 4/1996
Schedule VI (*See Rule 12*)

I................................wife/daughter of ....aged ....years residing at present at ....with permanent address as given below, do hereby state on solemn affirmation that I have been explained fully (in English/Hindi/ Marathi or in the language that I understand) the probable side effects and the after effects of the pre-natal screening test. I wish to undergo the procedures in my interest viz. With a view to find out the possibility of deformity or disorder, etc. In the child, which I am likely to deliver.

I undertake that I shall not terminate the pregnancy if the diagnosis shows the possibility of a normal child with either male or female sex. I understand that sex of the fetus will not be disclosed to me.

I understand that breach of this undertaking will make me liable to penalty as prescribed in the Maharashtra regulation of use of pre-natal diagnostic techniques Act, 1988.

Date: **Signature**

ANNEXURE 2: First Trimester Screening Test Report Proforma

SRL DIAG, Dr Phadke's Pathology
210, Ground Floor, Rambaugh, LJ Road,
Mahim, Mumbai–400016

Prisca	5.0.2.37
Date of report:	20/06/14

Patient data

Name		Patient ID	10
Birthday	20/10/87	Sample ID	10
Age at sample date	26.7	Sample Date	18/06/14
Gestational age	12 + 0		

Correction factors

Fetuses	1	IVF	no	Previous trisomy 21	unknown
Weight	87	diabetes	no	pregnancies	
Smoker	no	Origin	Asia		

Biochemical data

Parameter	Value	Corr. MoM
PAPP-A	1.25 mlU/ml	0.66
fβ-hCG	16.7 ng/ml	0.44

Risk at sampling date

Age risk	1:863
Biochemical T21 risk	<1:10000
Combined trisomy 21 risk	<1:10000
Trisomy 13/18 + NT	<1:10000

Ultrasound data

Gestational age	11 + 2
Method	CRL Robinson
Scan date	13/06/14
Crown rump length in mm	47
Nuchal translucency MoM	0.78
Nasal bone	present
Sonographer	Dr Geeta Shah
Qualification in measuring NT	MD, DMRD

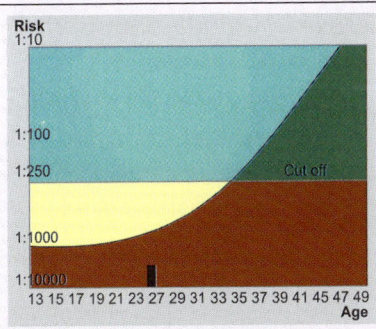

Trisomy 21

The calculated risk for Trisomy 21 (with nuchal translucency) is below the cut-off, which indicates a low risk.

After the result of the Trisomy 21 test (with NT), it is expected that among more than 10000 women with the same data, there is one woman with a trisomy 21 pregnancy.

The calculated risk by PRISCA depends on the accuracy of the information provided by the referring physician.

Please note that risk calculations are statistical approaches and have no diagnostic value! The patient combined risk presumes the NT measurement was done according to accepted guidelines [Prenat Diagn 18:511–523 (1998)]. The laboratory cannot be hold responsible for their impact on the risk assessment! Calculated risks have not diagnostic value!

Trisomy 13/18 + NT

The calculated risk for trisomy 13/18 (with nuchal translucency) is <1:10000, which

represents a low risk.

Sign of Physician

ANNEXURE 3: Guidelines for First Trimester Report Interpretation

1. The above test is a screening test only for trisomy 13, 18 and 21 and none other. It is not a definitive diagnostic test.

2. The interpretation should be analyzed in conjunction with all other factors, tests and clinical findings judged relevant.

3. Please note that only biochemical markers PAPP-A, free β-hCG without NT have a detection rate of only 60%, while inclusion of NT increases that to 85% at 5% false positivity.

Recommendations

a. *For patients detected in high risk zone*: (Risk >1:100, when cut-off is 1:250) at term

   1. Confirmatory test by *chorion villus sampling (CVS) and orkaryotype studies* is recommended.

   2. If the patient/clinician does not wish to undergo invasive confirmatory tests of CVS, in 1st trimester, then it is recommended that *triple test* along with *anomaly scan* to be done at 18 weeks and *amniocentesis* can be performed for confirmation of diagnosis, if needed.

   3. A second trimester screening test (triple marker) is recommended as *combined test (1st and 2nd trimester)* give over 80% detection rate (at 5% false positive rate).

b. *For patients detected in low risk zone*: (Risk <1:500, when cut-off is 1:250)

   1. This is not a definitive diagnostic test and low risk does not exclude the possibility of Down's syndrome/ Edwards as detection rate is 85–90% with 5% false positive rate.

   2. Detailed anomaly scan is recommended at 18–20 weeks along with triple screen test for neural tube defect (NTD) and Down's syndrome.

 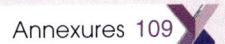

c. *Patients detected in borderline zone*: (Risk 1:100 to 1:500, when cut-off is 1:250)

1. This requires careful counseling, if there is any other risk factor like advanced maternal age, diabetes, ultrasound marker, family history of Down's syndrome then immediate diagnostic testing by CVS may be necessary.

2. In absence of any such high risk factors an anomaly scan at 18–20 weeks and second trimester serum screening may be recommended.

ANNEXURE 4: Second Trimester Triple/Quadruple Test Sample Referral, Patient Consent Proforma

Note: * Mandatorily to be filled

*Name: *DOB:

Address:

*Religion: *Caste: Age: Weight:

*Clinician incharge:

*E-mail ID:

*Hospital address:

*Contact numbers:

Diabetes: Yes / borderline /No   Smoking: Yes /No

Periconceptional folic acid: Yes/No

Brief obstetric history:

*Gestational age by LMP: *LMP

Sonography findings: *Scan date

Fetal status: Single / multiple =

Gestational age: *BPD:

History/off Down's/neural tube defect/other anomalies:

Was the first trimester screen test performed ? What was the result?

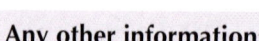

Any other information:

Instructions for sample collection:

1. For second trimester screening (UE3, hAFP, β-hCG and inhibin A) the sample should be collected between 15th to 21st week of pregnancy.

2. A 5CC of blood must be collected in a plain bulb.

3. Serum can be stored for 6 – days at 2–8°C.

4. Transportation should be done at 2–8°C. (In ice box)

Contact:

*Consent form for Prenatal Screening Tests
Schedule VI (*See Rule 12*)

I.................................….. wife/daughter of ....

Aged............years residing at present at ....with permanent address as given below, do hereby state on solemn affirmation that I have been explained fully (in English/Hindi/ Marathi or in the language that I understand) the probable side effects and the after effects of the pre-natal screening test. I wish to undergo the procedures in my interest viz. With a view to find out the possibility of deformity or disorder, etc. In the child, which I am likely to deliver.

I undertake that I shall not terminate the pregnancy if the diagnosis shows the possibility of a normal child with either male or female sex. I understand that sex of the fetus will not be disclosed to me.

I understand that breach of this undertaking will make me liable to penalty as prescribed in the Maharashtra regulation of use of pre-natal diagnostic techniques Act, 1988.

Signature

SRL DIAG, Dr Phadke's Pathology
210, Ground Floor, Rambaugh, LJ Road,
Mahim, Mumbai–400016

Date of report:		20/06/14
Prisca		5.0.2.37

Patient data		**Ultrasound data**	
Name		Gestational age	18 + 3
DOB	07/10/83	Scan date	09/06/14
Age of delivery	31.1	Method	Scan

Correction factors

Fetuses	1	IVF	no	Previous trisomy 21	unknown
Weight in kg	69	diabetes	no	pregnancies	
Smoker	no	Origin	Asian		

Risk at term

Age risk at term	1:843	Trisomy 21	1:1788
Overall population risk	1:600	Trisomy 18	<1:10000
Neural tube defects risk	<1:10000		

Pregnancy data		**Parameter**	**Value**	**Corr. MoM**
Sample date	18/06/14	AFP	30.73 ng/ml	0.49
Gestational age at sample date	19 + 5	hCG	9587 mlU/ml	0.60
		uE3	1.31 ng/ml	0.90
Determination method	Scan			

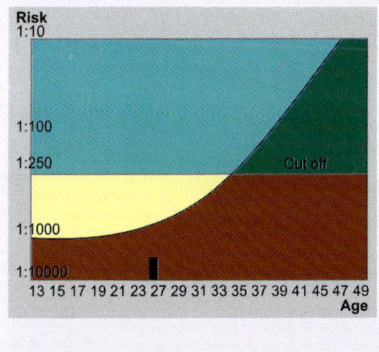

Trisomy 21

The calculated risk for Trisomy 21 is below the cut off which represents a low risk.

After the result of the Trisomy 21 test, it is expected that among 1788 women with the same data, there is one woman with a trisomy 21 pregnancy and 1787 women with not affected pregnancies.

The calculated risk by PRISCA depends on the accuracy of the information provided by the referring physician.

Please note that risk calculations are statistical approaches and have no diagnostic value!

Trisomy 18	**Neural tube defects risk**
The calculated risk for trisomy 18 is <1:10000, which indicates a low risk.	The corected MoM AFP (0.49) is located in the low-risk area for neural tube defects.

ANNEXURE 5: Guidelines for Second Trimester Screening Reports Interpretations

The above test is a screening test only for trisomy 18 and 21 and neural tube defect (NTD) and none other. It is not a definitive diagnostic test.

The interpretation should be analyzed in conjunction with all other factors, tests and clinical findings judged relevant.

Recommendations for Downs Syndrome Screening

a. *For patients detected in high risk zone:* (>1:200)
  1. The definitive/confirmatory test is amniocentesis and chromosomal studies.
  2. High-risk patients are suggested to undergo amniocentesis for karyotyping studies, as practiced all over the world.

b. *For patients detected in low risk zone:* (<1:300)
  1. This is not a definitive diagnostic test and low risk does not exclude the possibility of Down's syndrome/Edwards in exceptional cases.
  2. Detailed anomaly scan is recommended at 18–20 weeks along with other relevant tests (as indicated).

c. *For patients detected in borderline zone:* (1:200 to 1:300)
  1. If there is family history/obstetric history/advanced maternal age then diagnostic test may be recommended.
  2. In absence of the above an anomaly scan at 18–20 weeks as well as fetal echocardiography should be suggested.

Index